C0-AWY-745

Energy Costs, Urban Development, and Housing

ANTHONY DOWNS AND KATHARINE L. BRADBURY
Editors

Energy Costs, Urban Development, and Housing

THE BROOKINGS INSTITUTION
Washington, D.C.

Library of Congress Cataloging in Publication data:

Main entry under title:

Energy costs, urban development, and housing.

Includes bibliographical references and index.

1. Housing policy—United States—Congresses.
2. Energy policy—United States—Congresses. 3. Urban
policy—United States—Congresses. 4. Inflation
(Finance)—United States—Effect of energy costs on—
Congresses. 5. Poor—United States—Energy assistance—
Congresses. I. Downs, Anthony. II. Bradbury,
Katharine L. III. Brookings Institution.
HD7293.E58 1984 333.79'0973 83-46033
ISBN 0-8157-1050-X
ISBN 0-8157-1049-6 (pbk.)

This work was prepared in part under Department of Housing and Urban Development Cooperative Agreement H-5248CA. The United States Government reserves a royalty free, nonexclusive and irrevocable license to reproduce, publish or otherwise use, and to authorize others to use, for United States Government purposes, all copyrighted material resulting from the work performed under this cooperative agreement.

9 8 7 6 5 4 3 2 1

THE BROOKINGS INSTITUTION is an independent organization devoted to nonpartisan research, education, and publication in economics, government, foreign policy, and the social sciences generally. Its principal purposes are to aid in the development of sound public policies and to promote public understanding of issues of national importance.

The Institution was founded on December 8, 1927, to merge the activities of the Institute for Government Research, founded in 1916, the Institute of Economics, founded in 1922, and the Robert Brookings Graduate School of Economics and Government, founded in 1924.

The Board of Trustees is responsible for the general administration of the Institution, while the immediate direction of the policies, program, and staff is vested in the President, assisted by an advisory committee of the officers and staff. The by-laws of the Institution state: "It is the function of the Trustees to make possible the conduct of scientific research, and publication, under the most favorable conditions, and to safeguard the independence of the research staff in the pursuit of their studies and in the publication of the results of such studies. It is not a part of their function to determine, control, or influence the conduct of particular investigations or the conclusions reached."

The President bears final responsibility for the decision to publish a manuscript as a Brookings book. In reaching his judgment on the competence, accuracy, and objectivity of each study, the President is advised by the director of the appropriate research program and weighs the views of a panel of expert outside readers who report to him in confidence on the quality of the work. Publication of a work signifies that it is deemed a competent treatment worthy of public consideration but does not imply endorsement of conclusions or recommendations.

The Institution maintains its position of neutrality on issues of public policy in order to safeguard the intellectual freedom of the staff. Hence interpretations or conclusions in Brookings publications should be understood to be solely those of the authors and should not be attributed to the Institution, to its trustees, officers, or other staff members, or to the organizations that support its research.

Foreword

STEEP increases in energy prices during the 1970s strongly influenced urban development and housing, though not always in ways most people expected. Many planners thought that higher energy costs would have a centralizing effect on urban areas. They reasoned that higher commuting costs would induce people to live closer to city jobs and that the higher heating and cooling costs of suburban single-family houses would make city apartments more attractive. Yet the 1980 census revealed that the population of all central cities grew by less than 1 percent from 1970 to 1980, whereas that of all suburbs combined grew by 18 percent in the same period. Thus costlier energy did not seem to have fostered urban centralization. And what—if anything—the city governments or the federal government should do about higher energy prices was not at all obvious.

Even so, Congress and the federal administration were strongly urged to help households and organizations adjust. Pressure was applied in the belief that the outcome determined by the free play of market forces alone would not be adequate and hence that public policies that would promote energy conservation and help some people pay higher energy costs were needed. Various proposals—for assisting poor households with heating bills and for requiring that all new structures embody specific energy-saving requirements, for example—were advanced. But the wisdom of such proposals depends greatly on how well households, businesses, nonprofit organizations, and public agencies are likely to adapt to higher energy costs *without* added government action.

Thus formulation of public policies suitably responsive to higher energy costs requires a factually and analytically sound basis for judging how those costs are already affecting urban development and housing. To construct such a basis, the U.S. Department of Housing and Urban Development asked the Brookings Institution to hold two conferences. The first, held in November 1980, defined assumptions about future cost and availability of energy in preparation for the second, held in November 1981. At the second conference, academic experts presented six papers assessing the effects of higher energy costs on various aspects of urban development and housing, including industrial location and regional development. Energy experts from public agencies commented on the papers and took part in the ensuing discussion, as did other specialists in energy, housing, and urban development.

This volume presents revised and edited versions of the six papers, commentary on them, and an overview by the organizers of the second conference, Anthony Downs and Katharine L. Bradbury. Downs is a senior fellow in the Economic Studies program at Brookings; Bradbury, a former Brookings research associate, is an economist at the Federal Reserve Bank of Boston. The other conference participants are listed on pages 287–88.

The editors and contributors are grateful for comments by members of the staff of the Department of Housing and Urban Development, especially David Engel. David Howell Jones edited the manuscript, Judith Cameron verified its factual content, and Nancy Snyder proofread it. Ward & Silvan prepared the index. Anita G. Whitlock circulated the original papers and handled the conference arrangements; Jacquelyn Sanks circulated the revised papers and otherwise followed the project to completion.

Preparation of the papers, arrangements for the two conferences, and the publication of this volume were financially supported under a cooperative agreement with the Department of Housing and Urban Development. The views expressed here are solely those of the contributors and should not be ascribed to the Department of Housing and Urban Development, or to the trustees, officers, or other staff members of the Brookings Institution.

BRUCE K. MACLAURY
President

November 1983
Washington, D.C.

Contents

Tables

Figures

ANTHONY DOWNS AND KATHARINE L. BRADBURY

What Does It All Mean?

HIGHER energy prices imply many things concerning future housing and urban development policies, but not necessarily those most often cited by the press and urban experts. Here we shall summarize the findings and policy implications that emerged from the Brookings Conference on Energy Prices, Housing, and Urban Development, held in November 1981. We shall offer a brief summary of the principal policy findings, summaries of the six papers presented at the conference, and a discussion of public policy issues that emerged from the papers and comment.

Summary of the Principal Policy Findings

Six general policy conclusions emerged from this conference.

Market mechanisms responsive to the rise in energy prices since 1973 are generating appropriate adjustments concerning most location decisions and other long-run allocations of resources. Higher energy prices will cause few households, firms, or nonprofit organizations to change their locations. This is true both within and among U.S. metropolitan areas, even in the long run. Higher energy prices have had, and will have, much greater effects upon the characteristics of the national housing inventory. Reliance on market forces to generate appropriate adjustments, however, is proving quite effective for single-family and owner-occupied housing. No major policy changes seem to be needed, therefore, to compensate for market failures concerning the long-run allocation of resources geographically or in the nonrental housing stock.

Some groups, however, are adapting less well to higher energy prices.

Policies to help them make the necessary adaptations would improve the overall economic efficiency of the country, reducing energy use by means that would pay for themselves. Some groups do not know what adjustments to make. These include many homeowners who need energy audits. Others lack the resources to make optimal adaptations. These include some fiscally hard-pressed local governments and school boards. Still others are prevented from making appropriate adjustments by specific market obstacles. The owners and tenants of most multifamily rental units, for example, do not have any incentive to conserve energy to the maximum degree. Some of these inhibitions to proper adaptation could be reduced through policy interventions in housing markets.

Many households are suffering from adverse income effects imposed by higher energy prices, even after making appropriate adjustments to those prices. Policy interventions to aid the most deprived of these households would be focused on issues of equity rather than on improvements in economic efficiency. Such interventions would redistribute income to households that failed to attain decent standards of living after paying higher energy prices. It would probably be more equitable to aid the poorest households, however, regardless of how they were affected by energy price increases, than to focus aid on those households injured most by such increases.

In the long run, entire energy-importing regions may also suffer adverse income effects from higher energy prices that are not felt in energy-exporting regions. Large increases in energy prices have shifted the terms of trade between energy-importing and energy-exporting states in favor of the latter. In addition, large state revenues from severance taxes and other energy-related sources may permit energy-exporting states to reduce their other taxes to relatively low levels. Low general taxes plus high energy accessibility will make these states attractive locations for economic activity. Such regional income effects will probably accelerate long-run migration into energy-exporting regions from energy-importing regions. It would be unwise, however, to try to offset these long-run effects through public policies to redistribute incomes among regions deliberately.

In theory, significant changes in future patterns of housing and urban development beyond those that will come about through market forces could help the country save energy. But there are more readily feasible ways to conserve energy than using government regulations to create such drastic changes. Massive alterations that would increase urban densities could reduce future energy use significantly. But they would require huge outlays of capital, and many other highly valued goals would have to be sacrificed. Other energy-conserving adjustments, moreover, such as the use of more fuel-efficient vehicles, can save just as much energy and cause far less social disruption.

If so-called energy crises occur in the future, strong political pressures will be exerted on the federal government to "do something" in response. It would be prudent to explore the likely effectiveness of various policies in advance to avoid taking ill-conceived actions in such situations. It is not certain that there will be acute energy shortages or sharp price rises in the future. But experience indicates that, if such crises emerge, many efforts will be made, both within and outside the federal government, to make significant changes in housing and urban development policies ostensibly for the purpose of conserving energy. Some of the actions that would be suggested would be ineffective or even harmful, but these and other inefficient actions that would bring local political benefits might be adopted anyway if persuasive proof of their unsuitability could not be presented quickly when an energy crisis arose.

Summaries of the Six Conference Papers

In three conference papers the probable effects of energy price increases on the locations of households and firms were investigated. Such effects might be expected if some locations provided access to less expensive energy than others, or some locations allowed firms or households to consume less energy while enjoying the same profits or standards of living as others. The three papers suggest that neither of these effects is very large; higher energy prices have thus not caused and are not likely to cause much change in location patterns.

How Energy Price Increases Affect the Location of Economic Activity

Roger Schmenner examines the various means through which rising energy prices might affect the location of industry. One way industry can respond to higher energy costs is by substituting other resources in production processes to reduce the amount of energy used per unit of output. Another involves changes in location, either to obtain locally cheaper energy or to reduce the need for energy in transporting plant inputs or outputs.

Schmenner's findings imply that for several reasons changes in location are not likely to be the primary response of industry to energy price rises. First, many manufacturers have begun to use energy-saving technologies in existing plants. Some of these technological changes are new, some embody existing conservation procedures that higher prices have made advantageous. U.S. industry actually used less energy in 1977 than in 1971. Second, although increases in the prices of oil, gas, and electricity have been substantial, price differences among states have narrowed considerably during the last two decades. The energy-price advantages of certain states and regions have therefore declined. Third, for most industries, differences in the cost of energy between one area and another are less important than other cost differences or controlling concerns. Decisions to relocate are not reached quickly, and the list of factors generally considered includes several not related to energy, such as labor costs, unionization, and the quality of life. Fourth, those few industries for which energy costs are crucial were already located where energy supplies and prices were most advantageous for their needs or as near as possible to their markets—for inputs or outputs—if their principal use of energy was for transport.

Regarding the third and fourth items, more research is needed to determine what proportion of firms—and of what type—are near enough to the margin of decision that an energy price increase would alter their preferences for certain locations. Schmenner's findings, however, imply that the proportion is not large. Even uncertainty about the future availability of energy seems to affect the fuel choices of firms and their decisions to adopt flexible technologies more than their choices of location.

William Miernyk discusses the effects of energy price changes on economic activity in regions within the United States. His thesis is that the regional income effects of recent developments in the energy market have been greater than the regional growth effects on population and employment. This argument is based on a large shift in the terms of trade between energy-surplus and energy-deficit states as energy prices rose during the 1970s.

Miernyk presents extensive data on energy production by type of fuel and energy consumption by type of fuel and by sector for the ten federal census regions. Comparing the British thermal units (Btu) of energy produced and consumed in each state indicates that thirty-seven states—

and six of the ten regions—have energy deficits. Louisiana and Texas—and the Southwest region in which they are located—have the largest energy surpluses. Each of these two states produces 12 percentage points more of the national energy supply than it consumes. Recent changes in energy prices have affected surplus and deficit areas differently. Energy prices are usually lower in producing states than in consuming states, but most interstate disparities in fuel prices have narrowed since the early 1970s. Even if prices were the same in all regions, however, producing states would gain from consuming states because of the overall rise in energy prices in relation to the prices of other goods and services.

Miernyk also presents data on changes in regional employment classified by its likely vulnerability to changes in energy prices. Jobs in all three categories of vulnerability have disappeared in energy-deficit regions at about the same rate. This lack of differential response supports Schmenner's argument that shifts in industrial location that are directly caused by changing differences in energy prices are small.

Miernyk also points out that state severance taxes in energy-producing areas exact further transfers from deficit states to surplus states. In the short run, these taxes allow beneficiary states to increase services or reduce other taxes; in the long run, more people and jobs may be attracted to these areas because of their relatively low nonseverance taxes.

Richard Muth questions the widespread belief that recent dramatic increases in energy prices will produce a large-scale return to the city. While the basic direction of energy-cost effects is toward centralization, their size is negligible.

It is through commuting costs that changes in energy prices have their strongest effect on urban locations. Urban densities and prices per acre of land typically decline with distance from the central business district. Land prices decline with distance to offset the greater travel costs of commuting downtown. Densities decline because each household can afford more land where the cost per acre is less. Also, density in the central city has declined in relation to suburban density as real incomes have grown and transportation costs have declined. With the recent increases in energy costs and thus in commuting costs per mile, however, workers in the central business district would theoretically like to move in closer to save on travel. But how large is this effect? Though gasoline prices have approximately doubled in real terms since 1973, total commuting costs embody many other elements besides gasoline, including the imputed cost of time spent. Muth calculates, in fact, that total commuting costs per mile have risen less than 5 percent since 1973. This small increase, when annualized and compared to housing expenditures, has a centralizing effect approximately equal to the decentralizing effect of half a decade of growth in real income. Thus, the decentralization of U.S. metropolitan areas may be slowed somewhat, but will certainly not be reversed, by increases in energy prices.

Choices of housing space and type of structure are also affected by changes in energy prices. Concentrations of both smaller units and multiunit residential structures are generally greater in central cities than in their suburbs. These types of structure use less energy per unit than larger units or single-family homes. Increases in energy prices therefore make central city housing stocks more attractive. Energy accounts for about 12 percent of total housing expenditures, and the typical single-unit structure uses almost twice as much energy as a comparable unit in a structure having five or more units. Nevertheless, Muth calculates, each of these factors could account for only a small net increase in the demand for central city housing.

Thus, these authors all conclude that even very large increases in energy prices have only minimal effects on the location of economic activity. This lack of response is not caused by inertia, institutional barriers, or other failures to respond to economic incentives. Rather, locational responses are small mainly because other means of adapting to higher energy prices are more efficient. Also, interarea differences in energy prices have declined as energy prices have risen; so moving from one area to another cannot greatly reduce the need to make other adjustments to much higher energy costs. Although rising energy prices create incentives to minimize transportation costs, the existing dispersion of economic activity is so extensive that no particular direction of movement or migration will dominate the response. For these reasons, no public policy interventions concerning the location of economic activity—enhancing or reducing mobility, for example—are called for on efficiency grounds.

Increases in energy prices have large interarea effects on the relative well-being of citizens, however. Some state governments—and their taxpayers— have benefited greatly from severance taxes paid by all consumers of energy. Similarly, because of climate, urban layout, or the existing stock of physical capital, standards of living depend on more intensive use of energy in some areas than in others, so higher energy prices have reduced "real" income more in some areas than others. These income effects will be discussed in more detail later.

How Increases in Energy Prices Affect Housing

In the other three conference papers the probable effects of rising energy prices and energy shortages on the U.S. housing stock were examined. Higher energy prices raise the desired level of energy efficiency in housing units. But the long-lived nature of housing raises a question about the speed with which changes in energy efficiency will actually occur. An average of only 1 percent of the existing stock is withdrawn from use each year, and about 2 percent was built during the preceding year.[1] Thus, normal turnover will only gradually improve average energy efficiency. That will happen if higher energy prices motivate builders to make new units more efficient and the energy efficiency of those units withdrawn from the stock is below the average. In addition, the energy efficiency of existing units can be enhanced through changes in the behavior of occupants—turning down thermostats, using less air conditioning, and so on—or investments in new equipment, such as storm windows and new furnace burners.

Dwight Jaffee summarizes the available information about the relation between energy costs and new residential construction. The basic questions are the degree to which new housing units are being made more energy efficient and the effects of energy developments on the amount of new housing construction. Total new construction is affected by changes in both the relative attractiveness of new construction and existing units and the general conditions of the demand for and supply of housing. Jaffee shows that consumers are indeed concerned about the energy efficiency of new housing units and that builders are changing the characteristics of new units to address that concern.

Surveys by the National Association of Home Builders demonstrate the keen interest of potential home buyers in energy costs and the energy-use characteristics of new homes. Data from the U.S. Department of Energy indicate that new units have more insulation and other energy-

1. Annual averages, 1973–78. See U.S. Bureau of the Census, *Annual Housing Survey*, pt. A, *General Housing Characteristics* (Government Printing Office, various years).

saving features, such as electric heat pumps, than have older units. Including such features adds to the initial price of a new housing unit but reduces its future operating costs. The rising costs of energy inputs in construction and the added capital for energy-saving features have contributed to housing-price inflation during the past decade. But Jaffee concludes that higher debt-service costs have been much more important than higher energy costs in reducing the affordability of owning a house.

Jaffee also reports that during the late 1970s the average price of older housing units rose faster than the average price of new units, reducing the gap between them. This is the opposite of what would be expected if problems of affordability were dominated by energy considerations. Then, since improvements in energy efficiency have been made faster in new units than in older units, the prices of older units would decline in relation to those of new units. The fact that they have not done so confirms the observation that energy considerations have not been central to the pricing of houses.

During the second half of the 1970s, the average characteristics of new single-family units changed in several ways. Three changes involved moves toward greater use of energy per unit: an increasing proportion of new units contained central air conditioning, the average number of bathrooms rose, and the average number of square feet of floor area increased. Whether these moves are more energy efficient or less so depends on the additional quality and quantity of housing services produced per additional unit of energy. Several other changes had the effect of reducing average use of energy. These included a decreasing proportion of single-family starts since 1975 and an increasing proportion of attached units among all single-family units. Because energy is only one of many factors that influence the characteristics of new housing units, these data are difficult to interpret. Jaffee argues that consumers seem more willing to reduce their consumption of energy by investing in greater efficiency than by reducing the level of housing services they consume. Average household consumption of energy by region varies much less than regional differences in weather as measured by heating degree days and cooling degree days. This shows that energy-conservation measures are being widely and effectively used in areas in which weather conditions are adverse.

Kevin Neels and Michael Murray analyze the effects of changes in energy prices on the existing housing stock. As is true of new housing, there are two principal types of effect: changes in the average characteristics of older units through retrofitting and selective removal and changes in market conditions and hence the market value of older housing units.

Neels and Murray describe the availability and costliness of several energy-conservation options for both new and older units. There is naturally greater flexibility in incorporating energy-saving technologies in units not yet built as improvements become available. In addition, technologies available to both, such as wall insulation or storm windows, are more cheaply included in new units than retrofitted into older units. Thus, in general, higher energy prices give new construction a competitive edge over the existing stock in meeting the demand for housing.

As a result, buyers will not be willing to pay the higher energy costs entailed in using the stock of older houses, so the value of older units will decline in relation to that of new units as energy prices rise, reflecting reductions in future flows of net revenue of the former. Neels and Murray argue that this competition will permeate housing markets, affecting the net return on all existing units, even those not directly competing with newly built units.

Using data on rental properties from the Experimental Housing Allowance Program in Green Bay, Wisconsin, Neels and Murray simulated the effects on the existing housing stock of annual energy price increases of 2 or 4 percent during the next fifty years. They assumed that the stronger possibility of adapting new units to increases in energy costs would constrain the market prices of older units. They then calculated the decline in building incomes and values attributable to the shrinking gap between increases in revenue and increases in energy costs. Assuming annual increases in energy prices of 2 percent, by 2025 the price of energy would equal 360 percent of its 1973 value. Neels and Murray estimate that the constant-dollar price of housing services would then be equal to about 142 percent of its 1973 value. For a worst case, in which no energy-saving improvements could be made in the existing housing stock, the values in constant dollars of existing buildings would decline slowly to a level equal to about 70 percent of their 1973 level by 2025. Faster increases in energy prices or significant technological innovations that would allow new units to economize much more on energy would accelerate this decline in the real value of inflexible older units.

These worst-case projections exaggerate the probable negative effects on older buildings for two reasons. First, owners of older buildings can reduce their competitive disadvantage by making some energy-saving improvements. Second, the average energy efficiency of the older stock rises as the least efficient units are removed. Further simulations in which these factors have been taken into account indicate that future net incomes from existing buildings would decline in the short run as revenues were diverted into energy-saving investments. But property values would stabilize at higher levels in the long run.

Experience concerning the behavior of landlords during the original Organization of Petroleum Exporting Countries oil embargo implies that energy-saving improvements are indeed available for older buildings and can be financed out of property income. Yet when energy prices are steadily rising, technologically available energy-saving improvements may not be as cost-effective today as they will become tomorrow.

Raymond J. Struyk provides much more detail on adaptations of the existing housing stock to rising energy prices, focusing on the energy expenditures and weatherization activity of low-income households and those headed by elderly persons. Struyk also examines the incentives to energy-saving improvements in the stock of rental housing, both subsidized and unsubsidized, and barriers to such improvements.

Low-income households generally occupy less energy-efficient housing and spend much larger shares of their incomes on heating and other uses of energy than higher-income households. Owner-occupied units are generally better weatherized than rental units. But since they are also larger, total energy expenditures for units occupied by owners and by higher-income households are greater.

The incidence of poverty among the elderly is higher than among the nonelderly, but the rates of homeownership by the poor elderly are higher than by other poor households. Struyk finds that the energy use and weatherization activities of the elderly generally resemble those of otherwise similar households headed by younger persons. As energy prices rose during the 1970s, fewer improvements in weatherization were made in housing units occupied by poor households, including federally subsidized housing, than in others. Struyk argues that occupants of subsidized units should be given strong incentives to make more efficient use of energy.

From the few data available, it appears that privately owned multifamily rental housing also showed only gradual improvement in energy efficiency in response to rising prices. Incentives to conservation of energy are weakened by the division of control between tenants, who regulate energy-use factors that affect their ways of living, and landlords, who decide on structural factors, along with mixed practices regarding payment for energy use. In addition, landlords who rent to low-income households generally find it more difficult to obtain investment financing.

Struyk suggests that most of the owner-occupied stock will become well weatherized in the future, but the rental stock will become so much more slowly. Some of this progress will be attributable to removal of the least energy-efficient units, which are often less desirable on other grounds as well. In spite of added weatherization, however, low-income households will continue to pay relatively large shares of their incomes for energy.

All three authors agree that the energy efficiency of the U.S. housing stock has improved in response to the energy price increases of the past decade. They also expect it will yet improve further. Housing units added to the stock each year through new construction are, before construction, the most flexible; they thus reflect the most appropriate levels of improvement. Within the existing stock, owner-occupied units have made the greatest progress in conservation. It seems that homeowners are more willing to spend money for energy-saving improvements, however, than to sacrifice other amenities, such as floor space, to reduce their consumption of energy. Since owner-occupied units are typically both single-family units and larger than rental units, they use more energy per unit in spite of their greater energy efficiency.

Because units not yet built are more adaptable than existing units, recent increases in energy prices have reduced the relative profitability of the existing inventory. This competition creates incentives to energy-saving investments in the existing stock, but also imposes capital losses on its owners.

The evidence suggests that, with a few exceptions, producers and owners of housing units are responding sensibly to higher energy prices. The exceptions suggest three possible avenues of public intervention. The energy inefficiency of federally subsidized housing calls for certain alterations in reimbursement rules, funding distinctions, or subsidy levels. These would create incentives—or remove disincentives—for owners or operators to make economically justified conservation investments. The slow response of private multifamily rental structures suggests a need for similar incentives that would affect both tenants and owners. Finally, the large share of income spent on energy by poor households, while not necessarily producing inefficiency, may call for some type of income-redistributing intervention on equity grounds.

Discussion of the Principal Policy Findings

In spite of the general effectiveness of market-driven adjustments to higher energy prices, some groups are failing to make such adjustments for several reasons. These are lack of information, specific obstacles to efficient operation of the market, and lack of the necessary resources. Certain public interventions could often enhance their responses to market forces, thereby improving the overall economic efficiency of the country.

The residential sector as a whole has responded to higher energy prices at about the same speed as other sectors. As all sectors reduced their use of energy, residential consumption of energy remained about 20 percent of the total from 1973 to 1980.[2] Even so, many households have failed to apply known energy-saving techniques to their own homes. To some extent that failure is attributable to lack of information.

Some observers believe households to be "economically irrational" if they do not quickly adopt every energy-saving improvement that reduces energy costs more than it costs to install. But refraining from doing so is quite sensible under certain circumstances. Other uses of available resources—such as earning interest from money-market funds—may produce even higher rates of return than energy-saving improvements. Many households plan to move before such improvements in their houses have time to pay for themselves. True, if those investments become capitalized in the values of their houses, they will recover their costs when the houses are sold. But they may not think it is worth the bother of installing improvements if they plan to move soon. Even uncertainty about the length of time they plan to remain in their present homes can cause such postponements. Also, if they live in deteriorating neighborhoods in which house values are declining, they may be unable to recover investments in energy-saving improvements when they sell.

2. See Sam H. Schurr and others, *Energy in America's Future: The Choices before Us* (Johns Hopkins University Press for Resources for the Future, 1979); Bureau of the Census, *Statistical Abstract of the United States, 1982–83* (GPO, 1982), p. 573.

Moreover, there is great uncertainty about future energy prices. The uncertainty was magnified in early 1982 by a temporary worldwide surplus of oil and a resultant decline in fuel prices in many areas. In addition, more efficient fuel-saving technologies may soon be developed, so some consumers hesitate to invest in present technologies. Thus, failure to install available technology need not spring from either inadequate information or economic irrationality.

Making Adequate Information Available to Households

Nevertheless, some people have failed to adjust to higher energy prices mainly because they did not know what to do. During the early 1970s, for example, potential buyers of new houses did not know how much energy houses of various designs would use, how much each use would cost, or how to reduce such costs. Yet rising energy prices made them more aware of the need to keep home energy costs low. Their sensitivity was strengthened by widespread publicity that was given to energy costs and conservation during the so-called energy crises of the 1970s and by rapid increases in their heating bills. Also, builders competing for business began publicizing the desirability of having energy-efficient elements in new houses and using such elements more often. This further increased acceptance by home buyers of energy efficiency as an important criterion in choosing a new unit. Thus, private market forces expanded the amount of information about energy conservation that was available to buyers of new houses.

In northern Ohio, for example, limitations on supplies of natural gas during the early 1970s forced many builders to switch to electric heating, even though it had not been common in that market.[3] Until then, electrically heated homes had a reputation for costly utility bills. To overcome this marketing handicap, some builders incorporated many more energy-saving features—such as heavy insulation and double-paned windows—into their new units than they had before. As a result, one builder reduced heating bills in his new electrically heated houses to a level well below those of nearby gas-heated units. Because of such practices, the average energy consumption of new homes sold in the

3. This case history was supplied at the conference by Robert Schmitt, of the National Association of Home Builders.

entire country fell about 50 percent in just ten years.[4] Thus, when the market provides good information to consumers and the prices they are charged represent true social costs, no further public intervention is necessary.

Households that need information about retrofitting existing homes to save energy, however, are less likely to get it from market forces alone. Many occupants need energy audits conducted by trained specialists to determine the best ways to make their homes more energy efficient. But these households are often uncertain whom they should ask to make such audits or install any desirable improvements. Elderly householders are especially wary of letting strangers enter their homes. Trustworthy local sources of information about persons who could be reliably employed for these purposes would be extremely helpful.

One possible source of such information is public utility companies. Wherever consumption of electricity approaches the generation capacity of local public utilities, they have strong incentives to help their customers reduce their use of energy so that they can avoid the costly and difficult process of increasing their generation capacity. They therefore have good reason to help their customers conduct energy audits as a step toward reducing consumption. In some areas, utilities not only publicize such audits, but conduct them, pay for them, or support other organizations that conduct them. Examples are programs of the Potomac Electric Power Company (PEPCO) in the Washington, D.C., area and the Mass-Save organization sponsored by fifty-three local gas and electric utilities in the Boston area.[5] If local utilities have excess capacity, however, they are likely to be less interested in helping their customers reduce energy use. The participation of utility companies in household energy conservation is thus most likely in areas in which there is rapid growth, heavy reliance on electric heating and air conditioning, or some other cause of energy shortages.

4. It normally takes at least fourteen years, and often much longer, for technical innovations to become widely used. See Real Estate Research Corporation, Public Affairs Counseling Division, "Factors Involved in the Transfer of Innovations: A Summary and Organization of the Literature," paper prepared for the U.S. Department of Housing and Urban Development, Office of Policy Development and Research, January 1976.

5. PEPCO offered a free do-it-yourself energy audit booklet or a professional audit for a low fee ($25). Mass-Save is a nonprofit organization formed by Massachusetts electric and gas utilities to provide energy conservation services to their residential customers. For a nominal fee ($10) it provided an in-home energy audit. Such audits made specific suggestions for energy-conserving measures and estimates of first-year savings and payback periods and furnished lists of firms and suppliers.

Households changing residences or looking for their first homes in an area need information about past energy use in the units they are considering in order to make the most energy-efficient choices. Sellers of older homes are usually willing to make their utility bills available. But where tenants pay for the utilities of a rental unit, potential renters often find it hard to obtain information from past tenants, who have usually departed. In some communities, potential renters can obtain such data from local public utilities, but in other areas there are laws that forbid such disclosure. Removal of these laws would make valuable information more accessible.

Some oil companies have objected to the information-gathering functions that the federal Energy Information Administration has required them to perform. These functions include reporting present and projected availability of fuel, rates of consumption, and prices, by specific fuel and by state. These data are needed, however, if state governments are to carry out the responsibilities concerning the use of energy that the federal government has delegated to them. President Reagan vetoed legislation that would have given him emergency fuel-allocation powers, so state governments would have to perform any such allocations appropriate in periods of acute fuel shortage.[6] State-level data would also help state governments plan the use of funds made available to them under the energy block grants being considered by Congress. In addition, such data are essential to any analysis of the regional effects of proposed legislation, such as an import tax on petroleum. It would be desirable, therefore, to maintain these reporting requirements.[7]

Institutional Obstacles to Appropriate Adjustments

Institutional obstacles constitute a second cause of the failure of certain groups to adjust appropriately to higher energy prices. Although some of these obstacles and their effects can be identified, no one knows what all of them are or how widespread their effects may be. These subjects should be a primary focus of future energy-related research.

6. Some critics of past emergency allocations believe private markets could handle "crisis" situations better than any government allocation schemes. This is why President Reagan vetoed federal emergency allocation powers.

7. See Gordon L. Weil, director, Maine Office of Energy Resources, "Testimony on Behalf of the National Governors' Association," speech given before the Subcommittee on Fossil and Synthetic Fuels of the House Committee on Energy and Commerce, March 9, 1982.

One obstacle is the inability of small-scale landlords to borrow enough capital to make energy-saving improvements in their rental properties. Banks are often reluctant to make such loans, because they do not want to accept rental units as security. Also, banks require shorter payback periods than are economical for some energy-saving improvements, such as the purchase of new furnaces. In areas where utility companies are close to their existing generating capacities, they might be willing to make such loans, as discussed earlier. Another remedy would be state or federal guarantees of such loans. These guarantees would probably not require any direct subsidies, but would help make markets more efficient by reducing the risks of private lending. Thus, they would perform the same function for energy-conserving improvements that Federal Housing Administration (FHA) loans have performed for homeownership since the 1930s.

A second obstacle is the unwillingness of small-scale owners to pay for retrofitting by refinancing their properties as long as interest rates are as high as they were from 1980 to 1982. Many properties were originally financed with fixed-payment mortgages at relatively low interest rates. The cash flows provided by these properties would be wiped out if the owners had to refinance them at much higher present interest rates in order to pay for energy-saving improvements. Cash flows from many rental apartment buildings in particular have already been squeezed by rapid increases in energy costs combined with slow increases in rents. A careful analysis of the economics of retrofitting older buildings of all types was made by the Office of Technology Assessment, on the basis of interviews with ninety-six property owners across the country. The conclusion reached was that some subsidies to reduce the costs of financing and the risks of investing in retrofitting might be necessary to encourage such investment by small-scale owners who must use debt financing.[8]

The biggest institutional obstacle to adaptation of the housing inventory to higher energy prices arises from the incentive arrangements inherent in most rental units. The amount of energy consumed in any housing unit is influenced by two factors: the behavior of the occupants and the design and structure of the unit and its principal energy-using equipment, such as its furnace. In rental units, the first factor is controlled by the

8. See Office of Technology Assessment, "Will Building Owners Invest in the Energy Efficiency of City Buildings?" in *Energy Efficiency of Buildings in Cities* (GPO, 1982), pp. 99–140.

tenants, while the second factor is controlled by the owners, who are responsible for any capital investments that alter the energy efficiency of the units. In most rental units, all utility bills for the units themselves—or at least the heating bills, which are usually the largest—are paid either entirely by the tenants or entirely by the owner. The owner normally pays the costs of utilities used in any common spaces, such as lobbies, garages, and hallways. For the most part this dichotomous arrangement creates positive energy-saving incentives for only one of the two actors that influence energy use: whichever one pays the utility bills.

If the tenants pay those bills, they are motivated to adopt energy-conserving behavior when energy costs rise. They can keep the windows and doors closed more of the time and set their thermostats lower in winter and higher in summer. But they cannot directly affect the physical structure of their units or the heating and cooling equipment, which also influences the amount of energy they use, since that is controlled by the owner. Tenants could make some physical improvements that would save energy, such as installation of storm windows. But such improvements become the property of the owner when the tenants move, so few tenants make them. Yet the owner has only a moderate incentive to invest in improvement of the energy efficiency of the unit, since he or she does not pay the utility bills. True, a unit whose energy costs are high is less competitive in the rental market with similar units that are more energy efficient, so its rent tends to be lower. The possibility of receiving higher rents gives landlords at least some reason to make energy-saving capital improvements.[9]

When the owner pays the energy bills, he or she has a strong incentive to make the unit physically more energy efficient by adding insulation, installing a better furnace, and so on, but the owner cannot directly influence those aspects of the tenants' behavior that affect energy use. And in this instance, the tenants have little motivation to conserve energy, since they do not suffer from higher utility bills.

The problem has many facets. In public housing, where local authorities usually—but not always—pay utility costs, they receive subsidies from the federal government to assist with operating costs. Housing authorities are fully reimbursed for higher energy costs caused by rate increases, so they are not motivated to save energy in response to such increases. But they are only reimbursed in part for higher costs caused by increases in

9. We are indebted to Kevin Neels for pointing out this fact.

consumption, and they are allowed to keep some of any savings that are the result of lower consumption. Hence they have some incentive to reduce energy use in their units. They have little control, however, over either the behavior of tenants or the funds available for physical improvements.

Until now, the U.S. Department of Housing and Urban Development (HUD) has not placed great emphasis upon improving the energy efficiency of the nearly 3.5 million housing units it directly subsidizes. Yet those units constitute more than 12 percent of all the occupied rental inventory in the country.[10] HUD blames this failure on tight budgetary restraints and higher-priority uses of its limited funds for improving subsidized housing. Even so, HUD should consider making those investments that would pay for themselves in reductions in energy costs. It should at least cause local housing authorities to face the full incentives of higher energy costs and allow them to use funds earmarked for physical improvements for energy-saving investments. This situation is discussed further in the paper by Raymond Struyk.

Another way to overcome this institutional obstacle would be to involve third-party investors where tenants pay for utilities, or where owners pay for them but cannot secure loans for energy-saving improvements. A third party motivated by potential profits might finance energy-saving investments through the required repayment period of, say, five to ten years. Where the tenants pay for utilities, the third party would contract with them to recover the capital costs as the savings materialized. Where owners pay for utilities, similar contracts would be made with them. Such arrangements would relieve both tenants and owners of the necessity of making the initial capital investment. Third parties could operate on a larger scale than individual tenants or owners could, thereby obtaining better access to capital and to qualified contractors and suppliers. They could also spread over larger numbers of projects the risk that any one improvement would not work.

Such third-party arrangements have serious drawbacks in practice, however. If tenants entered into long-term contracts for improvements, they would be unable to move as soon as they might otherwise without carrying the debt with them. Average annual tenant turnover in U.S.

10. Jill Khadduri and Raymond J. Struyk, "The Case for Housing Vouchers in a Time of Fiscal Restraint" (Urban Institute, June 1981), p. 2.

rental housing markets is about 37 percent a year, and it is higher in the South and West.[11] The total savings available from making most rental units more energy-efficient, moreover, are too small to motivate investors to grapple with the risks and complexities of many small-scale projects. True, third-party arrangements have worked in projects to improve the energy efficiency of some industrial buildings. But that was primarily because of much larger potential savings from improving each structure and also because of substantial federal tax incentives.

The only third parties likely to invest in improving the energy efficiency of much U.S. rental housing are public utility companies, which have additional motives for doing so. As noted earlier, the strength of those motives will vary among utility companies, depending upon how close each is to full utilization of capacity. The success of third-party arrangements is therefore likely to differ greatly in different parts of the country. Even so, such arrangements are not likely to cause a significant improvement in the energy efficiency of the rental housing stock.

Another way of reducing this institutional obstacle would be to induce both tenants and owners to share equally in paying utility bills. Then both would be equally motivated to conserve energy. They could undertake concerted efforts, combining modifications in tenant behavior with investments by owners in physical improvements. Local governments should therefore explore the desirability of requiring such shared-utility-cost arrangements in all future residential leases involving buildings in which each unit has separately metered utilities. In buildings without separately metered units, this shared-cost approach would not work.[12] Such a radical step might require revising many existing arrangements between tenants and owners, including changes in levels of rents that do not include utilities. In return for that disruption, the new arrangement would create stronger combined incentives for the principal actors to reduce energy consumption in the many U.S. rental units that do have

11. Bureau of the Census, *Annual Housing Survey: 1980*, pt. D, *Housing Characteristics of Recent Movers for the United States and Regions* (GPO, 1982), p. 1.

12. Owners must pay all utility bills in such buildings, so to share the costs, they would have to allocate those overall bills to individual tenants. But each tenant would claim to be conserving energy mightily, blaming any waste on all the others. Without separate metering, the owner would have no way of verifying these claims; hence the owner could not effectively link each tenant's utility bills to that tenant's energy-usage behavior. We are indebted to Marshall Holleb for pointing out this important limitation to the utility-cost-sharing approach.

separately metered utilities. Perhaps the most effective way to explore this policy would be through a federally sponsored demonstration in one or two cities.

Aiding Those with Inadequate Resources

Some people lack the resources to adjust appropriately to higher energy prices. In particular, they cannot substitute other factors for energy in response to its higher price. Their situation is similar, but not identical, to that of persons injured by the adverse income effects of higher energy costs.

These groups would like to reduce their consumption of energy, but they cannot afford the required steps. Examples are poor homeowners who want to weatherize their energy-inefficient homes, fiscally strapped school boards that want to install more efficient furnaces and better insulation in their old schools, and equally destitute local governments that want to modernize their cavernous, energy-wasting public buildings. Their plight is especially acute if they are located in cold areas, such as the Northeast, where heating fuels are also the most costly.

From the viewpoint of national efficiency, it would be desirable to help such groups substitute other factors for energy. But such assistance would inherently involve some redistribution of incomes, thereby raising issues of equity as well as efficiency. The equity aspects of such aid will be discussed in the next section; here we shall focus on the most efficient way of providing it.

There are four basic ways to furnish these groups with federal or state assistance in response to their higher energy costs. One is to offset increases in their energy costs with assistance tied directly to those increases. This is what some energy-assistance-payment programs do for low-income households. This method is inefficient, however, because it reduces the incentives of the recipients to conserve energy, since any increases in their energy costs are at least partially passed on to taxpayers. If the aid covers only a fraction of their energy-cost increases, they will have at least some incentive to reduce their use of energy, but not as much as if they had to pay the full market price of energy.

A second approach is to tie aid to installation of weatherization or other energy-saving improvements. This reduces the overall consumption of energy in the country; it is therefore more efficient than the first approach. Examples are weatherization grants or loan programs for owner-occupants and local governments and tax credits for installation of energy-saving improvements. Such programs are described briefly in the paper by Raymond Struyk.

The third approach is to provide federal or state funds to households in relation to their poverty or some other criterion not tied to energy at all, but permitting those funds to be used for conservation of energy at the discretion of the household. Recipients of such aid would have more purchasing power generally, some of which they would use in substituting other resources for energy. An example is allowing local governments to use their Community Development Block Grants for energy-conservation purposes if they so desire. This approach would improve the energy efficiency of U.S. buildings less per public dollar spent then the second approach but more than the first one.

The fourth approach is to lend these homeowners and groups the money to make those improvements that are cost-effective enough to pay for themselves in time, presumably at low interest rates. This is administratively difficult, but it would reduce the total long-run cost of such aid to taxpayers.

Adverse Household Income Effects of Higher Energy Prices

Because energy prices have risen much faster than the general price level since 1973, their increase has imposed significant losses in real income upon all households and organizations except owners of energy resources. These adverse income effects occur when households or organizations must spend a larger share of their available resources than formerly to buy the same amount of energy or less. As a result, they have less purchasing power with which to buy other goods and services. The larger the share of total income a household spends on energy and the greater the price increase, the severer such income effects are. Some of these adverse effects can be escaped if the household responds to the higher price of energy by using less of it. The household can substitute other resources for energy, as when it buys insulation, can change its habits to conserve energy, as when it lowers its thermostats during winter, or can simply use less, as when it closes off part of its home. But purchased energy performs certain vital functions, such as heating and cooling homes, providing light and electrical power, and powering vehicles, so most households cannot come close to reducing their use of energy in proportion to its recent price increases.[13] They must therefore spend a larger share of their incomes for energy than they did before its price rose, even if they consume less of it. And some forms of reduced energy consumption require a significant lowering of their true standards of living.

Whether or how public policy ought to respond to the resultant losses of real income is closely related to general social policies concerning poverty and the redistribution of income. After all, energy prices were not the only ones rising rapidly during the inflationary 1970s; other price increases also inflicted adverse real-income effects upon millions of households. Therefore, the income effects of energy price increases upon poor households should be analyzed in the light of what happened to their overall budgets.

Ideally, this analysis would cover the incomes of these households, including transfer payments and assistance in kind, the cost of heating and cooling their homes, the cost of other energy connected with their housing, their nonenergy housing costs, the cost of their food, what they spend for gasoline, their taxes, and their medical expenses. This would require detailed analysis of the budgets of low-income households at various times before and after recent energy price increases.

The closest thing to such comprehensive information that is readily available is the lower-budget urban household expenditures compiled each year by the Bureau of Labor Statistics (BLS). Table 1 shows data concerning changes in essential elements of such budgets for a four-person household for the periods 1970–81 and 1976–81. The eleven-year period encompasses both the major energy price increases of 1973–74 and 1979; the five-year period includes the 1979 increase. Total budgetary costs for such a household rose 120 percent during the entire eleven years and 53 percent during the last five years.

Four essential budgetary items have been labeled nonenergy items because their consumption by the household involves relatively little direct expenditure for energy—though production of these items, especially food, involves considerable expenditure for energy. These are food, medical care, social security taxes, and income taxes. Total spending for these four items amounted to 50.7 percent of the entire budget in 1970, 53.1 percent in 1976, and 56.2 percent in 1981. Two other essential budgetary items have been labeled energy items because their con-

13. In economic jargon, their demand for energy has a low price elasticity.

Table 1. *Changes in Essential Items in the Annual Budgetary Expenditures of an Urban, Lower-Budget, Four-Person Household, 1970–81*
1970 = 100

Item	*Increase, 1970–81*		*Increase, 1976–81*	
	Percent	*Dollars*	*Percent*	*Dollars*
Total budget	120.2	8,363	52.6	5,282
Nonenergy items	143.9	5,082	61.7	3,285
Food	138.6	2,640	51.3	1,542
Medical care	155.5	874	60.3	540
Social Security taxes	200.3	691	71.5	432
Income taxes	122.0	877	93.5	771
Energy items	113.4	2,194	51.2	1,397
Housing	97.1	1,388	43.4	853
Transportation	159.6	806	70.9	544

Source: Bureau of the Census, *Statistical Abstract of the United States, 1972* (GPO, 1972), p. 350; ibid., *1977* (1977), p. 485; ibid., *1982–83* (1982), p. 465.

sumption involves considerable direct expenditure for energy and represents most of the expenditure for energy by the household. These are housing and transportation. Total spending for these two items combined amounted to 27.8 percent of the entire budget in 1970, 27.2 percent in 1976, and 26.9 percent in 1981. Spending for housing and transportation also included many nonenergy elements.

Table 1 shows that prices of the four nonenergy items combined rose about 27 percent faster during the period 1970–81 than prices of the two energy items combined and 21 percent faster during the period 1976–81. The combined energy items did not increase as fast as the total budget in either period. In both periods, moreover, the absolute increases in spending by this hypothetical household for the nonenergy items were more than twice those for the energy items combined. In fact, the increase in the cost of food alone exceeded combined increases in the costs of housing and transportation in each period. Similar conclusions emerge when changes in the low-income budget of a two-person household some of whose members are sixty-five or over are analyzed.

Thus, *in absolute terms, the adverse income effects of increases in energy prices upon low-income households have been much smaller than those of increases in other parts of their budgets, both for the 1970s as a whole and for more recent years.* True, these data do not reflect the indirect effects of higher energy prices, such as their effects on food prices. Some analysts believe that indirect energy costs are equal to as much as 65–85 percent of direct energy costs.[14] No information is available, however, in which such indirect effects are accurately incorporated into budgetary analysis.

Another qualification is that the BLS low-income budget involves much larger total expenditures than persons officially defined as poor can afford. In 1981, the BLS low-income budget of a four-person urban household called for expenditures of $15,323. But the official poverty-level income for such a household was $9,287—or 39 percent less. Similar disparities existed concerning the BLS budgets for 1970 ($6,960 for low income versus a poverty-level income of $3,968) and 1976 ($10,041 for low income versus a poverty-level income of $5,815).[15] The conclusions about the effects of higher energy costs, therefore, may not apply to households whose incomes are at or below the poverty level. A 1974 analysis of low-income households, defined as those in the lowest 4 percent of the income distribution, showed that they spent 29.6 percent of their incomes on direct energy costs. This amounted to 13.4 percent of their total consumption spending, which exceeded their total incomes.[16] Another analysis indicated that a group with a median income of only $3,549 in 1979 spent 29.5 percent of its income on energy—9.9 percent for gasoline and transportation and 19.7 percent for household energy. The average direct expenditure of the U.S. median-income family for energy in 1979 was 11.2 percent of income.[17]

These data imply that truly poor households were hit harder by rising energy prices than the BLS low-budget analysis indicates, since they spent more of their incomes on energy than the households on whose expenditures the BLS data were based. This conclusion is especially likely for poor households living in cold regions. In 1978–79, spending on all residential fuels as a percentage of household income was 8 percent throughout the country, but it was 9 percent in the Northeast, 8 percent in the North Central region, 7 percent in the South, and 5 percent in the West.[18]

The effects of higher energy prices upon the poor also depend upon

14. See Hans H. Landsberg and Joseph M. Dukert, *High Energy Costs: Uneven, Unfair, Unavoidable?* (Johns Hopkins University Press for Resources for the Future, 1981), p. 33.

15. Bureau of the Census, *Statistical Abstract of the United States, 1982–83* (GPO, 1982), pp. 417, 465.

16. Landsberg and Dukert, *High Energy Costs*, p. 35.

17. Ibid., p. 32.

18. Ibid., p. 48.

the rapidity with which their total incomes have risen in relation to energy prices. Their incomes include earned income, transfer payments, and assistance in kind. Accurate data concerning changes in all these elements are difficult to obtain. The median money income of families whose incomes are below the poverty level rose from $2,524 in 1973 to $4,478 in 1980, or 77.4 percent.[19] The median money income of unrelated individuals whose incomes are below the poverty level increased from $1,411 in 1973 to $2,826 in 1980, or 100.3 percent.[20] During the same seven years, the consumer price index for energy increased 232 percent.[21] Although these data on money incomes do not include assistance in kind, energy prices clearly rose much faster than the real incomes of the poor during the 1970s. This undoubtedly inflicted serious losses in real income on poor households who had to continue spending large shares of their incomes for energy. Most seriously injured were poor households living in cold climates and using fuel oil for heating, since natural gas prices were controlled. When natural gas prices are fully decontrolled, poor households living in cold areas and using natural gas for heating will probably suffer similar losses in real income.

Policy Responses to Adverse Household Income Effects

Should society try to offset the adverse income effects inflicted on the poor by rising energy prices? That depends partly on whether antipoverty assistance should be concentrated on people whose income levels are low or people whose incomes have declined greatly. The losses of income inflicted by rising energy costs are declines in real income suffered by everyone who must continue to use energy. Poor households are less well able to bear such declines, or any other increases in cost, than other households for the obvious reason that they have fewer resources to

19. The number of such families was 5.5 million in 1975, fell to 5.3 million from 1976 through 1979, then rose to 6.9 million in 1981, but mainly because new data from the 1980 Census were then introduced. Hence there was probably no actual change in the number of such families during this period that would cause a notable shift in the median income.

20. The number of such individuals rose steadily during this period, from 5.1 million in 1975 to 5.7 million in 1979 and 6.2 million in 1980. Data on the number and incomes of families and persons below the poverty level from Bureau of the Census, *Current Population Reports*, series P-60, "Characteristics of the Population below the Poverty Level" (GPO), various issues.

21. Information obtained by telephone from the Bureau of Labor Statistics, Consumer Price Index Office.

begin with. If society wants antipoverty aid to respond to the overall relative deprivation of low-income households, however, it should allocate such aid in relation to income levels, not to changes in income levels.

Consider two households of the same size, age, and other primary characteristics except location and income. The Toasties live in a warm climate with an income of $2,000 a year, while the Chills live in a cold climate with an income of $5,000 a year. The Toasties spend only 10 percent of their income on energy, leaving them with $1,800 for all other consumption. In contrast, the Chills spend 20 percent of their income on energy, leaving them with $4,000 for all other items. If energy prices doubled, each household would reduce its energy consumption by 10 percent. But it still would have to pay much more for energy than before, leaving it with less to spend on all other items. The Toasties would cut their energy consumption, measured in preincrease terms, from $200 to $180, but would then have to pay $360 for that amount of energy. They would thus sustain a net loss of $160 in purchasing power for other goods and services, retaining $1,640 of such purchasing power. Since the Chills spent much more on energy initially, they would suffer a much larger loss of purchasing power for other things when energy prices rose. Their net loss would be $800, or five times that of the Toasties. Even in relative terms, the Chills would be hit harder, losing 16 percent of their initial real income, while the Toasties would lose only 8 percent. Yet the Chills would still have a nonenergy purchasing power of $3,200, almost twice that of the Toasties. The incomes of both these households are below the poverty level, but which is in greater need of assistance from society?

Most economists would argue that the Toasties are in greater need, since their real income is much lower than that of the Chills, both before and after the energy price increase. Therefore, if society is going to aid the poor, it should aid the Toasties more than the Chills, even though energy price increases injured the Chills much more. This implies that aid to the poor should *not* be proportional to, or even related to, the effect of energy-price increases on them. Rather, it should be related to their general level of poverty, regardless of what caused that poverty.

True, if society had already been aiding both the Toasties and the Chills in proportion to their real income levels before the energy price rise and sought to aid them similarly afterwards, the Chills might receive a larger increase in aid than the Toasties, since the Chills suffered a larger decline in real income. Nevertheless, the Toasties would receive more aid, both before and after the energy price increase, since their real income was lower in both instances.

Some analysts might argue that what makes people suffer most, at least in the short run, is any drastic negative change in whatever they have become accustomed to. If such a change is permanent, those affected by it may need temporary assistance while adjusting to their new situation. Society can therefore legitimately offer interim aid to people whose real incomes have suddenly declined sharply because of nationwide forces beyond their own control, even if their incomes are not extremely low. By this reasoning, since the Chills have suffered a greater loss in real income than the Toasties, the Chills "deserve" more temporary compensation because of the energy price increase. This reasoning, however, does not justify permanently extending greater aid to the Chills because of the greater decline in their real income. Yet Congress sometimes appears to uphold the latter approach, since it often responds to changes in conditions, rather than to long-established levels of conditions. Its support of home-heating assistance to households injured by energy-price increases is an example.[22] Nevertheless, it is less equitable to provide long-term aid in response to changes in people's conditions than in response to the levels of their conditions.

Most federal antipoverty programs fail to take into account variations in the cost of living across the country. Such variations are especially relevant to energy considerations, since both energy use and fuel prices differ greatly from one area to another. Hence programs designed to aid the poor generally might nevertheless recognize regional variations in the cost of living that are related to energy use. This recognition should *not* consist of tying antipoverty aid directly to each household's specific expenditures for energy. That would greatly weaken the recipients' incentives to conserve energy. But to include some variations in the cost of living attributable to regional differences in energy costs and use would be sensible. This variable element might be overall housing costs, which are among the most regionally diverse elements of consumer budgets across the country, partly because they include the cost of home energy. In that case, those who administer the antipoverty assistance program should be sure their measure of housing costs includes appropriate energy components.[23]

22. If recent home-heating assistance is viewed as only temporary aid to help people adjust to higher energy prices, it can be justified through the "interim aid" argument mentioned above. Whether that justification applies in this case depends upon the length of time Congress continues such aid.

23. Some housing-cost measures are based on rents exclusive of utilities. Such measures would not be appropriate for this use.

This analysis also implies that federal funds now being used for home-heating assistance would be more equitably deployed if they were used for general antipoverty assistance. In fact, the $1.78 billion spent for federal home-heating assistance in 1981 could have provided a substantial part of the total cost of a nationwide housing voucher program for renter households whose incomes were equal to less than 50 percent of the median for their respective areas.[24] Jill Khadduri and Raymond Struyk estimate that such a housing voucher program, begun in 1982, would gradually reach the point of serving 2.5 million households in 1987 at an annual cost of $5.9 billion in 1980 dollars. It would include a regionally variable cost-of-living component.[25] True, it would not aid owner-occupants, and many poor householders—especially elderly ones—own their own homes.

Nevertheless, Congress would create a much more equitable and efficient program of antipoverty assistance if it combined energy and emergency assistance and housing vouchers into a single nationwide program. It would be more equitable because greater funding would permit it to aid a larger share of low-income households than the present energy-assistance program. It would be more efficient because much more of the assistance in a housing-voucher program is equivalent to a general income supplement. Such a supplement would provide greater benefits to the recipients—from their viewpoint—per dollar of public funding than aid tied to specific consumption, because it would place fewer constraints on the use of the aid.

Some observers believe, however, that many separate programs that tie antipoverty aid to specific types of consumption, such as food stamps and medicare, provide more total benefits to the poor than would a single general income-support program.[26] They contend that Congress is willing to vote much greater total funding for several earmarked aid

24. Federal outlays for energy and emergency assistance actually amounted to $1.78 billion in 1981 and $1.875 billion in 1982; the estimated outlay for 1983 is $1.986 billion. The Reagan administration has requested outlays of $1.3 billion a year in current dollars from fiscal 1984 through fiscal 1986. *Budget of the United States Government, Fiscal Year 1983*, p. 5-144; ibid., *Fiscal Year 1984*, p. 5-113.

25. Khadduri and Struyk, "The Case for Housing Vouchers, p. 18.

26. True, each dollar spent on transfers in kind is worth less to the recipients at the margin than a dollar of income transfer, because the latter does not constrain them to buy items they do not want. The willingness of Congress to spend more on transfers in kind thus does not necessarily make the poor better off. Whether they are better off depends partly upon how much more Congress will spend on transfers in kind and how they will be distributed.

programs than for one general income-assistance program, for three reasons. First, aid tied to consumption of specific goods or services often benefits the producers of those items.[27] The producers therefore offer strong political support for such aid that they would not offer for general income assistance. Second, many taxpayers want antipoverty aid restricted to specific types of consumption of which they approve. They do not want to permit "irresponsible" recipients to "squander" their aid on activities the taxpayers consider frivolous or harmful, such as gambling and liquor. In fact, there is little evidence that many recipients of general income support use their aid in these ways. But that fact is almost irrelevant if many taxpayers fear that they might. Third, the fragmented committee structure of Congress encourages the maintenance of many specific kinds of aid. Doing so enhances the political power and influence of more members of Congress than would consolidation of antipoverty aid under a single committee.

It can therefore be argued that maintaining a separate energy-consumption-assistance program is better for the poor than merging such aid into more general income assistance, including a housing-voucher program, even though the former is less efficient economically. Underlying this argument is the assumption that such aid for energy consumption is part of a larger set of several earmarked assistance programs. To decide whether this argument is correct would require making a political judgment that is beyond the scope of this book.

Regional Income Effects of Higher Energy Prices

Elsewhere in this volume, William Miernyk argues that higher energy prices cause the energy-exporting regions of the United States to enjoy large gains in income at the expense of certain energy-importing regions. If so, in the long run, more people and jobs will migrate to the income-gaining regions from the income-losing regions than would have done so otherwise. Should public policies respond to this situation?

Most economists, including most of those who attended the conference, believe that people and resources should be able to respond to regional differences in costs and benefits by moving freely around the country as

27. Such aid does not *always* benefit those producers because it may be provided in forms that are almost fully convertible into cash, as were food stamps. In that case, earmarked aid does not aid producers any more than general income assistance would.

they wish. Therefore, trying to prevent interregional movements that arise because of differences in energy prices or incomes is not an appropriate goal of public policy. Nor should society try to offset these income effects fully by redistributing incomes among regions.

True, regional migrations of people and jobs induced by energy prices might intensify poverty, unemployment, or urban decay in certain areas. It would then be appropriate to aid residents or governments of those areas through national programs in response to such adverse conditions wherever they were found, regardless of their causes. Thus, national public policies should not have the aim of influencing regional conditions as such, but should respond to particular adverse conditions regardless of where they occur.

Nevertheless, it might be desirable to distribute nationally some of the revenues obtainable from taxing sources of energy at their points of production or extraction, rather than confining those revenues solely or mainly to the states in which those points are located. Such revenues could be shared nationally through special federal taxes on energy sources. That is a highly controversial subject, however, that cannot be fully considered in this book.[28]

Pressures to Make Radical Changes during "Energy Crises"

Experience during 1973–74 and 1979 shows that government agencies are usually pressured to do unwise things whenever a so-called energy crisis appears. Both private citizens and public officials are likely to advocate radical changes in housing and urban-development practices in the effort to reduce energy use. The people who suggest these changes focus their attention almost entirely upon possible savings in energy, ignoring the high costs of attaining them.

Such pressures will always arise during energy crises for two reasons. First, activities associated with housing and urban development account for a large share of total national energy use. About 20 percent of all consumption of energy in 1980 took place within housing units.[29] The geographic pattern of urban development, moreover, immensely influences the amount of energy used for transportation. Since automobiles

28. William Miernyk comments on this subject in his brief analysis of state energy severance taxes.

29. Bureau of the Census, *Statistical Abstract of the United States, 1982–83*, p. 574.

absorbed more than 10 percent of total national energy use in 1980, anything that reduces their movement could significantly reduce total consumption of energy.[30]

Second, there is persuasive evidence that these types of energy consumption could be reduced dramatically. That would happen if techniques already known were intensively employed in modifying and operating existing housing, building new housing, and changing residential densities so as to alter patterns of urban transportation. Energy use within most existing housing could be reduced 25 percent or more through better insulation, more efficient furnaces, and greater efforts at conservation on the part of residents. Energy use in newly built units could be reduced even more by combining these tactics with better site placement, more south-facing windows, and passive solar designs.

If the country could switch rapidly from its present low-density pattern of urban settlement, requiring extensive automobile travel, to higher-density settlements and greater reliance on public transportation, the use of energy for transportation could be reduced dramatically. But such a switch is impossible, since the low-density pattern is already embedded in concrete, steel, bricks, and mortar across the country. Hence efforts to change urban density must be directed mainly at future growth. That drastically reduces the scope for potential savings in energy. Nevertheless, even marginal changes in urban development patterns could reduce somewhat the amount of energy used for transportation.[31]

For these reasons, activities associated with housing and urban development make tempting targets for people eager to reduce energy consumption during periods of shortage or for the long run.

Actually changing those activities so as to produce large savings in energy, however, is extremely difficult. An important reason is the asymmetrical relation between energy use on the one hand and housing and urban development on the other. Although energy use is greatly influenced by housing and urban development, housing and urban development are not greatly influenced by energy use—as is pointed out in several of the papers in this volume. With some exceptions, energy forms a relatively small part of the total cost of building and operating both housing and urban settlements generally. There are many ways for households and developers to adjust to higher energy prices, moreover,

30. See ibid.; and Schurr and others, *Energy in America's Future*, p. 75.

31. See Katharine L. Bradbury, Anthony Downs, and Kenneth A. Small, *Urban Decline and the Future of American Cities* (Brookings Institution, 1982), pp. 217–36.

besides altering the basic low-density settlement pattern of the country. Most people greatly prefer those other adjustments to high residential density. Therefore, it is neither economically rewarding nor necessary for them to make extensive changes in the existing settlement pattern in response to higher energy prices.

As a result, most households, urban governments, and home builders and other real estate developers resist making extensive changes in established ways of creating and operating urban settlements in order to save energy. True, they will adjust their behavior in marginal ways to reduce energy costs. Examples are adding insulation to dwellings, changing thermostat settings, and shifting to more fuel-efficient vehicles. These changes save a great deal of energy, but not nearly as much as more radical changes in density of settlement would. Yet the latter would also impose high costs upon households, governments, and developers, without providing them with commensurate benefits. An important conclusion of this conference is that their resistance to radical changes in patterns of settlement is quite consistent with the actual costs and benefits involved, in spite of higher energy costs. Therefore, few radical changes in density of settlement aimed at saving energy are likely to be made.

This discussion may seem academic, since future energy crises appear to be less likely in the light of the worldwide surplus of oil in early 1982. But many plausible events could precipitate another energy crisis. For example, a serious disruption of oil production in Saudi Arabia could create acute oil shortages and skyrocketing prices like those that occurred in 1979 when production in Iran fell sharply. Even rapid increases in decontrolled natural gas prices within the United States could create a domestic energy crisis. In either situation, elected officials and public interest groups would exert strong pressure on the federal government to "do something" immediately.

Past and Future Responses to Pressures for Change

In past energy crises, HUD was pressured to carry out many changes that would affect housing and urban development. Some involved alteration of existing regulations to place greater emphasis upon conservation of energy. As a result of such pressures in the 1979 energy crisis, HUD changed 200 pages of regulations in a single morning. Most of these changes consisted of the insertion of language that made conservation of energy a goal in dozens of activities in which it had until then been ignored. Examples are public housing programs, the Urban Development Action Grant program, and the Community Development Block Grant program. These alterations also permitted the use of HUD funds for comprehensive energy planning.

Other pressures arose for new regulations requiring more stringent energy-conservation standards in all new buildings. In response, the Department of Energy drafted a set of Building Energy Preservation Standards. But many people in the housing and building industries opposed these regulations as both unnecessary and too costly. Consequently, after eighteen months of debate, they were dropped. Other groups wanted a federal requirement that every existing housing unit be brought up to high-level energy-conservation standards before it could be sold. But bringing this idea to fulfillment involved many difficulties. They included the impracticality of using a single energy-conservation standard throughout the country, the diversity of local code-enforcement practices, the long delays likely to be caused in sales, and, in many instances, the economic infeasibility of making the required investments. This effort, therefore, was also rejected.

But several important measures were adopted as a result of such pressures. These included tax credits for energy-conserving improvements in single-family homes, federal financial aid to low-income households subjected to large increases in heating costs, and a weatherization-assistance program. The federal government has spent billions of dollars on these measures both directly and indirectly, since it has lost some revenue because of tax credits. Their effectiveness is discussed in the papers by Dwight Jaffee and Raymond Struyk.

Another program now being considered involves the use of federal funds to stimulate district heating and the cogeneration of heating and power in urban areas. Some mayors consider this a promising public works program that could pump construction funds into their depressed economies and provide badly needed jobs. Yet district heating schemes are not likely to be economically efficient.[32]

Nevertheless, these and other inefficient actions that produce local political benefits may be adopted anyway if persuasive proof of their unsuitability cannot be quickly presented when an energy crisis arises.

32. See Office of Technology Assessment, "Prospects for District Heating," in *Energy Efficiency of Buildings in Cities* (GPO, 1982), pp. 165–96.

As usual under such conditions, groups favoring certain policies for reasons unrelated to the crisis itself will advocate those policies as effective responses to that crisis. This is especially likely because the policies indicated at this conference as the most appropriate urban-related responses to high energy costs are not very popular politically. Many such policies involve redistribution of incomes to counteract the adverse income effects of higher energy costs upon poor households. Other appropriate policies would favor dissemination of better information to consumers about conservation of energy. Such policies absorb scarce resources of city governments without producing notable political benefits for local elected officials. In contrast, some policy responses judged inappropriate at this conference are likely to be much more acceptable politically, especially under the pressure of a sudden energy crisis. Many involve federal subsidies that would bolster big-city economies, although they are unnecessary or inefficient economically. Thus, the economic appropriateness of many urban policy responses to high energy costs appears to be incongruent with their political acceptability. This incongruency makes it urgent to analyze the suitability of various urban policy alternatives in advance of any energy crisis in order to avoid adopting inefficient policies when such a crisis arises.

It is not possible to catalog in this book all the ideas related to energy conservation that might be advanced under such circumstances and then analyze their desirability. Some were discussed earlier in this chapter. Others are analyzed in the remainder of the book. But many others are not evaluated here at all. It would therefore be prudent for HUD and other federal agencies that have energy-related responsibilities to identify as many such policies as possible and to conduct or commission careful studies of the advantages and disadvantages of each. Then, if pressures for hasty action should arise during future energy crises, persuasive arguments concerning the relative wisdom of such actions can readily be offered. This effort should certainly include evaluation of those energy-conservation policies adopted as a result of energy crises in the past.

RAYMOND J. STRUYK

Home Energy Costs and the Housing of the Poor and the Elderly

In a recent paper William Marcuse, summarizing several studies in which the share of income that households in the United States would devote to expenditures for residential energy in 1985 was compared to the share in 1975, reported that the disadvantage of the poor would increase.[1] What do such projections imply concerning the type of housing the poor will inhabit in a decade? Will the units be smaller and better weatherized but still not improved sufficiently to offset rising energy prices? How will the stock of housing itself be affected? Can widespread abandonment of energy-inefficient multifamily properties in severe climates be expected, for example? After surveying the available evidence, I find Marcuse's summary in general valid, although the path to the conclusion seems to involve a good deal of change in the extent of homeownership among low-income households and the type of housing they occupy. Obviously, if the type of housing occupied by the poor and the elderly is expected to change, there must be adjustments throughout the housing market.

The main conclusions of my paper are the following:

Weatherization. Owner-occupants live in housing that is substantially better weatherized than that of renters. Poor households consistently live in less well weatherized units, but in the coldest areas two thirds of the poor live in dwellings with some attic insulation and some storm

The author thanks Lorene Yap, Richard Morgenstern, Arthur Reiger, David Engel, Katharine L. Bradbury, and Anthony Downs for comments on a draft of this paper.

1. "Energy Pricing and Availability, 1980–1990," paper presented at a Brookings Institution conference, Housing and Energy in the 1980s, October 23, 1980.

doors and windows. On the average the units occupied by the elderly do not differ in the extent of weatherization from those occupied by others.

Home energy expenditures. Households with incomes of 75 percent or less of the poverty line spend a much higher share of their incomes on home energy even than households with incomes of 75–125 percent of poverty. No significant difference between elderly and nonelderly households is evident. The incidence of households with incomes below the poverty line that spend an extensive share of their income on home energy is about the same among the elderly and nonelderly, but about half of all households are elderly.

Adjustments to energy price increases. Short-term price elasticities for home energy use are inelastic overall but elastic for some uses, such as air conditioning. There is no evidence that the poor are more price-responsive than the nonpoor. Less weatherization is being installed in housing occupied by the poor—especially expensive items, such as new furnaces—than in housing occupied by the nonpoor; there is little distinction in this regard between units occupied by the elderly and the nonelderly. The poor may be abandoning the least efficient—and otherwise undesirable—units; those leaving the inventory are chiefly rental units in two- to four-unit and larger structures and those without central heating systems.

Public assistance with weatherization. The U.S. Department of Housing and Urban Development (HUD) now subsidizes about 3.5 million housing units but has done little to encourage improvements in weatherization or energy conservation in these units. The Weatherization Assistance Program (WAP) of the U.S. Department of Energy and federal tax credits have helped a large number of low-income, almost exclusively homeowning, households. By the end of 1981, 20 percent of the homeowners eligible for assistance in 1975 will have been helped. During the first twenty months of the availability of tax credits 171,000 households whose incomes in 1978 were less than $8,000 took credits. As an example of the amount of weatherization done in a more general home-improvement program in a subsidized rehabilitation loan program in Wisconsin, about a third of the improvements were found to involve weatherization.

Multifamily rental housing. There is very little information on this part of the housing stock. A higher share of low-income renters in such structures do *not* pay for their own utilities; their landlords should therefore have stronger incentives to weatherize. Landlords of low-income tenants, however, face greater disincentives to upgrade their properties than others because of greater cash-flow problems, difficulties of obtaining financing, and, in many instances, declining neighborhood conditions.

A look ten to fifteen years into the future. Homeowners are likely to be more affluent, and younger, because the elderly tend to shift from owner-occupied units at higher rates. The owner-occupied stock will almost universally be "adequately" weatherized. The exceptions will be in rural areas. More and more low-income households will be renters and be living in two- to four-unit and larger structures. Those paying especially high shares of their incomes for home energy will be those in units that today are individually metered or will be in declining neighborhoods and master metered. Because of the sluggishness of the low-income rental market, inefficient structures may remain in the occupied inventory for a surprisingly long time. The elderly as a distinct group will be unimportant; it will be their representation among the low-income renters in inefficient units that will matter most, and on this score their situation is very likely to improve during the years ahead.

The Housing-Energy Situation in the Late 1970s

In preparing this section I have relied exclusively on existing tabulations of data. Generally, these tabulations were prepared in conjunction with studies of specific issues. In many instances the cross-tabulations are not as detailed as would have been desirable for our purposes and the germane multivariate analysis has not been done. Consequently, while a general picture of the present situation has been developed, its resolution is not as sharp as it might be if further analysis of the same data sets were undertaken.

Some Basics

While the general profile of the low-income population is well known, a brief review will ensure a common starting point. Table 1 gives some basic facts for families and unrelated individuals whose incomes were less than 125 percent of the poverty level in 1978—$8,327 for a nonfarm family of four. (Poverty is defined in terms of cash income; thus the value of transfers in kind is ignored.) In total about 13 million primary families

Table 1. *Selected Characteristics of Primary Families and Primary Unrelated Individuals Whose Incomes Were Less than 125 Percent of the Poverty Level in 1978*

Characteristic	*Poverty households as percentage of all households*		*Poverty households with selected attributes as percentage of all poverty households*	
	Families	*Individuals*	*Families*	*Individuals*
Age of head				
Under 65	12.3	21.0	83.1	45.5
65 or older	14.5	43.4	16.9	54.5
With related children under 18	16.7	n.a.	72.3	n.a.
Tenure				
Owner-occupants	7.9	28.9	47.0	46.0
Renters	26.7	29.5	53.0	54.0
Location				
In metropolitan areas	11.2	25.8	58.9	64.2
In central cities	16.5	28.9	35.2	39.0
Outside central cities	7.6	22.2	23.7	25.2
Outside metropolitan areas	15.6	38.2	41.0	35.7

Source: U.S. Bureau of the Census, *Current Population Reports*, series P-60, no. 124, "Characteristics of the Population below the Poverty Level, 1978" (Government Printing Office, 1980), table 46, pp. 200–01.
n.a. Not available.

and unrelated individuals living alone—about 17 percent—were below this cutoff. The incidence of poverty among families headed by elderly persons is about the same as among other families, is about four times as great among renters as among homeowners, and is moderately higher in nonmetropolitan areas than in metropolitan areas. Among primary individuals, the elderly are twice as likely to be impoverished as the nonelderly, and those living in nonmetropolitan areas about one eighth more likely to be impoverished than those in metropolitan areas.

Tenure has important implications for the extent to which a unit is weatherized and for home energy expenditures. Shown in table 2 are the tenure distributions for 1977 of elderly and nonelderly households by income class (columns 1–3) and also the share of each tenure group composed of those over the age of 65 (columns 4–5). At the lowest income levels the elderly exhibit much higher rates of homeownership than their younger counterparts. Indeed, the elderly constitute an absolute majority of homeowners whose incomes are less than $7,000. Within this income group the incidence of homeownership among the

Table 2. *Tenure by Income and Age of Head of Household, 1977*

Annual income (dollars)	Rate of homeownership (percent)			Households headed by elderly as percentage of home-owners	Households headed by elderly as percentage of renters
	All households	Households headed by elderly	Other households		
Less than 3,000	43	53	37	48	32
3,000–4,999	47	61	32	65	36
5,000–6,999	51	73	38	54	21
7,000–9,999	53	78	44	39	12
10,000–12,499	58	83	54	21	6
12,500–14,999	66	84	64	16	6
15,000–19,999	72	85	71	10	4
20,000–24,999	81	86	80	6	4
25,000–34,000	86	83	87	5	6
35,000 and over	90	88	90	7	8

Source: Bureau of the Census, *Annual Housing Survey: 1977*, part C, *Financial Characteristics of the Housing Inventory* (GPO, 1979), table A-1, pp. 1–5.

elderly is about twice that of the nonelderly. The obverse holds among renters: two thirds of renters whose incomes are less than $7,000 are headed by nonelderly persons.

Weatherization

In describing the extent to which units are weatherized, my focus is solely on the presence of several easily measured features that help to make the envelope of a structure airtight. The available sources of data employ three measures—the presence of storm windows, of storm doors, and of attic insulation. The data used here from the Annual Housing Survey (AHS) of the U.S. Bureau of the Census simply record whether the unit has "some" of each of these items. In the other data set employed, the National Interim Energy Consumption Survey (NIECS), more precise information on each of these items and also on the presence of wall insulation was ascertained. The figures presented below are based on the AHS definition, unless otherwise noted.[2] It should be stressed that

2. For descriptions of the surveys, see John M. Goering, *Housing in America: The Characteristics and Uses of the Annual Housing Survey*, Annual Housing Survey Studies 6 (Government Printing Office, 1979); and U.S. Department of Energy, *Residential Energy Consumption Survey: Conservation*, DOE/EIA-0207/3 (GPO, 1980).

the available measures are far from ideal, providing only a rough indication of the extent of weatherization.

A number of factors are likely to influence the decision whether to weatherize a dwelling. The severity of the climate is an obvious factor. Whether the occupant owns the unit is another, since owner-occupants realize the savings from investments in weatherization. Among rental units in which tenants pay their own utility bills, the linkage between savings realization and investment is less direct—a subject to which I shall return later. The income of the resident is probably important, especially in the case of a homeowner, because it affects his ability to absorb the cost of improvements and possibly the rate at which he discounts future benefits in relation to present benefits. Age of the household also may be important in itself, because the elderly are likely to have higher discount rates and are as a group less likely to be able to make improvements themselves and, in some instances, are hesitant to engage contractors for fear of being cheated or of having strangers in their homes.

The examination begins with a quick look at the variation in weatherization with tenure and type of structure. It might first be noted that there is a fairly high correlation between a unit's being a single-unit structure, including a mobile home, and its being owner-occupied:[3]

Type of structure	*Percentage of units owner-occupied, 1979*
Single unit, detached	80
Single unit, attached	55
Mobile home	81
Structure having 2–4 units	21
Structure having 5 or more units	6

At least for single-family housing tenure distinction is very important: the incidence of having all three energy-saving components—storm windows, storm doors, and attic insulation—is about twice as great among owner-occupied units (52 percent) as among rented units (28 percent). The incidence of units having none of the components, moreover, is four times as large among rented units (28 percent) as among owner-occupied units.[4]

3. U.S. Bureau of the Census, *Annual Housing Survey: 1979*, pt. A, *General Housing Characteristics* (GPO, 1980), table A-1, p. 1.

4. Jeanne Goedert and others, "Character and Dimensions of the Need for Residential Energy Conservation," Working Paper 1344-1 (Urban Institute, 1979), table I.4A, p. 53.

Data on the incidence of energy-saving components by type of structure show single-family detached units to be the most nearly weathertight among structures having up to four units. Mobile homes and units in structures having two to four units have the least weatherization.[5] In the light of these facts, it may be of interest to note that, except for mobile homes, there is only modest variation in the income of renters with the type of structure they occupy.[6] Also, the elderly—both owners and renters—occupy single-family units at a lower rate than do younger households.[7]

As might be expected, there is a sharp variation in the extent of the weatherization according to the severity of the weather in the area. Table 3 shows the incidence of energy-saving features among single-family dwellings for four regions defined by Goedert and her colleagues on the basis of heating and cooling degree days in each state.[8] In the areas having the two coldest climates, all three components are present in three units out of four; in the two warmer areas, the incidence is less than one in three. Also, in the colder areas, practically no units are without any of the three energy-saving features.

How do the poor and the elderly fare in these areas? The incidence of energy-saving features in single-family units definitely rises with income. In the coldest area, however, two thirds of the units occupied by those whose incomes are below the poverty line have all three energy-saving components. The simple relation to income, of course, provides

5. Department of Energy, *Residential Energy Consumption Survey: Conservation*, table 4B, p. 20.

6. Median incomes of renters in 1977, by type of structure, according to the AHS:

Single-family unit	$9,100	Structure with 5–19 units	$8,900
Structure with 2–4 units	8,300	Structure with 20 or more units	9,300
Mobile home	7,000		

7. The 1976 structure distribution of households, by age of head of household, those sixty-five or over being classified as elderly, and tenure, in percent, was:

Number of units in structure	*Owner-occupants*		*Renters*	
	Elderly	*Nonelderly*	*Elderly*	*Nonelderly*
1	90	95	22	33
2–4	7	4	28	29
5 or more	3	1	50	38

See Raymond J. Struyk and Beth J. Soldo, *Improving the Elderly's Housing: A Key to Preserving the Nation's Housing Stock and Neighborhoods* (Ballinger, 1980), table 3-4, p. 42.

8. For a description of the survey, see Goedert and others, "Residential Energy Conservation," pp. 44–45.

Table 3. *Incidence of Energy-Saving Components among Single-Family Units,*[a] *by Climate, 1976*
Percent

Climate	Units having all components	Units lacking at least one component	Units lacking all three components
Total	49.1	50.9	9.9
Very cold	83.0	17.0	0.5[b]
Moderately cold	76.6	23.4	1.3
Warm	32.1	67.9	14.5
Hot	18.4	81.6	19.9

Source: Jeanne Goedert and others, "Character and Dimensions of the Need for Residential Energy Conservation," Working Paper 1344-1 (Urban Institute, 1979), table I.3a.
a. Mobile homes are excluded.
b. Based on ten observations or fewer.

only a rough indication of the income-weatherization relation for the poor because information on current income, as opposed to permanent income, is being used—an especially serious problem for the elderly in general and elderly homeowners in particular. Nevertheless, a positive income elasticity of weatherization appears to hold.[9]

Differences in incidence of energy-saving features associated with age are comparatively modest and provide little basis for arguing that the units occupied by the elderly are especially in need of weatherization.

Expenditures

In considering the outlays by households for energy consumed in conjunction with occupancy of their residences the distinction must be made between consumption and expenditures. For homeowners, the distinction is less important than for renters. Among renters who do not pay for their utilities separately, the problem of separating expenditures on energy from contract rents can be formidable. My principal interest in analyzing expenditures is to determine the extent to which the poor and the elderly are spending a disproportionate share of their incomes for energy. For this purpose it would be preferable to use data that imputed energy expenditures to renters who pay for utilities as part of their rent. Unfortunately, most of the figures employed here are for

9. Goedert and others, "Residential Energy Conservation," table I.32A, p. 101; notes to table I.27, pp. 91–92; table I.29A, p. 95.

Table 4. *Total Residential Expenditures for Energy, by Household Characteristic, April 1978–March 1979*
Dollars

Household characteristic	*Average expenditure*
All households	724
Census region	
Northeast	887
North Central	821
South	674
West	469
Type of structure	
1 unit, detached	800
1 unit, attached	742
Structure having 2–4 units	655
Structure having 5 or more units	476
Mobile home	587
Tenure	
Owned	805
Rented[a]	559
1977 family income (dollars)	
Under 5,000	522
5,000–9,999	627
10,000–14,999	659
15,000–19,999	769
20,000–24,999	816
25,000 or more	938

Source: Department of Energy, *Energy Consumption Survey: Consumption and Expenditures, April 1978 through March 1979*, DOE/EIA-0207/5 (GPO, 1980), table 1, pp. 12, 13, 15.
a. Units for which no cash rent is paid are excluded.

homeowners and for those renter households who pay their heating fuel bills separately.

A broad picture of home energy consumption is available from the published tabulations of the NIECS data. According to these data, in which an explicit imputation of utilities is made for rental units when it is necessary and figures reported by utility companies are used for electricity and natural gas consumption, the average household in 1977–78 spent $724, or 8 percent of its income, on home energy consumption (see table 4). Single-unit structures require more energy for heat and for other utilities, despite their higher degree of weatherization, although this varies with such factors as location and size of the unit and the income of the household. The average household in the Northeast spends

Table 5. *Percentage of Household Income Spent on Home Energy*

Type of energy and region	*Household income as a percent of poverty-level income*			
	Less than 75	*75–125*	*125–200*	*More than 200*
Heating fuels only				
Northeast	26	9	5	3
North Central	16	7	4	2
South	14	5	3	1
West	12	4	2	1
All home energy				
Northeast	41	16	10	5
North Central	29	14	8	4
South	30	13	8	4
West	25	9	5	3

Source: Wayne Lee Hoffman, "The Distribution of Home Energy Expenditures by American Households in 1976–77: An Analysis of Energy Need among Low-Income Groups," Working Paper 1197-3 (Urban Institute, 1979), table 6, p. 16; table 7, p. 18.

about 89 percent more on utilities than its counterpart in the West. On the average owners spend much more than renters. While home energy consumption increases steadily with income, the elasticity is certainly less than unity.[10] Unfortunately, I have not found data on the variation of expenditures with income after controlling for region, tenure, and type of structure.

Table 5 gives the essential information about the share of income devoted to utilities. Those households whose incomes are less than 75 percent of poverty-line income are spending a much greater share of their income for energy than even those in the group whose income is 75–125 percent of poverty-line income. Households whose income is less than 75 percent of poverty-line income in the North Central and South Census regions, for example, spend about 30 percent of their incomes for home energy and about half this for the fuel used in heating.

These points are reinforced by data not shown, in which the figures on the share of income spent on heating fuels are disaggregated by tenure and by age of the head of the household. There is a smaller variance in expenditure across income groups for energy used for heating fuels than

10. For a summary of estimated income elasticities, see Marcuse, "Energy Pricing and Availability," table 16, p. 37; table 17, p. 37.

in expenditures for other household activities.[11] Among households whose incomes are 75 percent of the poverty level or less, homeowners in the two cold regions spend a great deal more for heating fuel than do renters in the same areas. Among the other income groups, little difference is observed. Even among the poorest households no significant distinction is apparent between households headed by the elderly and other households.

Finally, it can be asked, for what share of households are energy expenditures a share of income higher than some "acceptable" threshold? Using two definitions of those in need of assistance with expenditures for home energy, Hoffman made such calculations for households living below the poverty line:[12]

Households in need of assistance with heating and fuel expenditures

Extreme need. Expenditures exceed 12 percent of annual income, or 25 percent of income for the six months of the heating season.

Moderate need. Expenditures exceed 6 percent of annual income, or 12 percent of income for the six months of the heating season.

Total home energy expenditures

Extreme need. Expenditures are at least 25 percent of annual income.

Moderate need. Expenditures are at least 14 percent of annual income.

While these criteria are arbitrary to some degree, they can be put into perspective by noting that throughout the country the average household spent 3 percent of its income on heating fuels and 7 percent on all home energy expenditures. Thus, to be in extreme need requires that a household be spending four times the average share of income on heating fuels and three times the average share on all home energy.

The figures in table 6 show the percentage of households with incomes of less than 125 percent of the poverty line, as defined by the Community Services Administration, that were in need according to these criteria. In the coldest region, 35 percent of these households are in extreme

11. This point is made by Harold Beebout, Gerald Peabody, and Pat Doyle, "The Distribution of Household Energy Expenditures and the Impact of High Prices," in Hans H. Landsberg, ed., *High Energy Costs: Assessing the Burden* (Johns Hopkins University Press for Resources for the Future, 1982), pp. 1–36; they point out particularly the much higher variance in expenditures for electricity across income groups.

12. Wayne Lee Hoffman, "Energy Need among Low-Income Households: State-Specific Estimates Using Several Alternative Measures," Working Paper 1197-5 (Urban Institute, 1979), pp. 9ff.

Table 6. *Poverty Households in Need During 1976–77, by Climate*[a]
Percent of all poverty households[b]

					Elderly-headed households	
	Expenditures for heating		Total expenditures for energy			
Climate	Extreme	Moderate	Extreme	Moderate	Percent in extreme need	Percent of all households in extreme need
Very cold	35	71	23	58	30	53
Moderately cold	28	66	20	54	24	55
Warm	25	60	18	51	24	54
Hot	22	54	20	54	22	53

Source: Wayne Lee Hoffman, "Energy Need among Low-Income Households: State Specific Estimates Using Several Alternative Measures," Working Paper 1197-5 (Urban Institute, 1979), pp. 11, 14, 29.

a. Figures are unweighted averages of the values for the states included as having a given climate.

b. Poverty households are defined as those whose incomes are less than 125 percent of the Community Services Administration (CSA) poverty line.

need of assistance with heating fuel expenditures and 23 percent are in extreme need of assistance with all energy expenditures. The incidence of need declines in more temperate climates, but even in the hot zone, 22 percent of households are in extreme need of assistance with expenditures for heating fuel.

The last two columns of the table give the share of the elderly in extreme need because of expenditures for fuel and the share of all households in extreme need that have elderly heads. The incidence of extreme need among the elderly is about the same as among other households. On the other hand, about half of all households in extreme need are elderly; directing utility assistance payments to the elderly may thus be a simple and fairly effective way to reach a substantial share of those in extreme need.

Adjustments to Rising Energy Prices: The General Picture

A family can respond to rising energy bills in many ways. It can turn its thermostat down, improve the weatherization of its unit, close off some rooms on especially cold days, turn the thermostat down further or use a space heater in one room, or move to a more energy-efficient unit. These few examples only suggest the range of responses. How have households—especially the elderly and those with low incomes—responded so far to rising energy prices since 1974? If these groups have been able to reduce home energy consumption sharply, it is clear that the extent of their plight has been greatly reduced. It will be apparent that the adjustments made by the poor have been modest—both absolutely and in relation to those made by the more affluent.

The information available is for fairly short periods or consists of statistical estimates based on cross-sectional data. Much of what is known about the types of adjustments to expect is therefore tentative. The discussion is organized into two blocks: short-term responses and long-term responses. The responses reviewed here are total—that is, both those induced by market forces and those encouraged by public action.

Short-Term Adjustments

Short-term adjustments can be defined as those that primarily involve changes in behavior rather than adjustments to the capital with which the household works to achieve an adequate level of energy-dependent services. There are some questions not answered by this definition: Does improved furnace maintenance done by a furnace company or the household constitute a change in the capital configuration? For simplicity, short-term adjustments are taken to include such simple and cheap adjustments to the existing capital.

Roberta Barnes and her colleagues have made estimates of the short-term expenditure and price elasticities for consumption of natural gas and electricity, using data from the 1972–73 Consumer Expenditure Survey of the Bureau of Labor Statistics and the rate schedules to which the households in the survey were subject.[13] Overall, the short-term response was found to be inelastic: −0.7 for natural gas and −0.5 for electricity. Separate estimates were also made for fuels for individual appliances; these are listed in table 7. This list of elasticities lower than −1.0 suggests that there are a substantial number of ways in which households can and do adjust consumption in response to higher prices.

The Annual Housing Survey provides some descriptive information on adjustments of this type. Table 8 shows the incidence of such adjustments in 1974 and in 1977 among households that lived in the

13. Roberta Barnes, Robert Gillingham, and Robert Hagemann, "The Short-Run Residential Demand for Natural Gas," *Energy Journal*, vol. 3 (January 1982), pp. 59–72.

Table 7. *Short-Term Price Elasticities of Demand for Fuels for Residential Appliances, by Type of Fuel*

Elasticity less than −1.0	*Elasticity greater than −1.0*
Natural gas	*Natural gas*
Central air conditioning	Space heater
	Water heater
Electricity	Range
Second refrigerator	Dryer
Dryer	
Portable air conditioner	*Electricity*
Space heater	Range
Central air conditioning	Freezer
	Dishwasher
	Color television
	Water heater
	Refrigerator

Source: Roberta Barnes, Robert Gillingham, and Robert Hagemann, "The Short-Run Residential Demand for Electricity," *Review of Economics and Statistics*, vol. 63 (November 1981), pp. 541–52.

same dwelling in both years. Separate figures are presented for owners and renters and for households headed by elderly and nonelderly persons, by income class. The three adjustments shown are: closing off rooms in the winter, reducing the number of breakdowns in heating equipment from the number suffered during the preceding winter—reflecting better maintenance but also possibly replacement of the heating plant—and using a secondary source of heat in addition to the main source. Among renting households no consistent pattern is revealed by the figures in the table. Among homeowners whose incomes were less than $5,000 in 1973, however, a larger share in 1977 than in 1974 were using additional sources of heat and had closed off rooms. This pattern held for those whose incomes were somewhat higher also. By contrast, the incidence of heating breakdowns increased among nonelderly homeowners, although it declined among elderly owner-occupants. On balance, though, these rough data are consistent with the extent of adjustments suggested by the elasticity estimates.

Long-Term Adjustments

As the adjustment period lengthens, a much broader array of choices, including improving the weatherization of their units and relocating to more efficient units, is obviously available to households. The long-term elasticities available suggest a decrease in consumption of no more than

Table 8. *Percentage of Households, Stationary during the Period 1974–77, That Made Adjustments in Their Heating Configurations, by Tenure, Income, and Type of Adjustment, 1974 and 1977*

Type of tenure, income, and year[a]	Additional source of heat[b]		Reduction in number of heating breakdowns[c]		Room closed off[d]	
	Non-elderly head	Elderly head	Non-elderly head	Elderly head	Non-elderly head	Elderly head
Renters						
Less than 5,000						
1974	8.5	7.0	1.2	0.8	8.4	10.4
1977	7.8	6.7	1.6	1.1	9.4	9.6
5,000–9,999						
1974	6.4	6.8	1.6	0.8	4.7	5.4
1977	7.2	7.1	1.5	1.1	5.7	5.1
10,000–14,999						
1974	5.4	5.3	1.1	0.7	2.8	4.0
1977	7.1	5.8	1.1	1.0	3.4	3.1
15,000–19,999						
1974	5.3	6.7	1.1	0.9	2.0	2.3
1977	5.4	4.8	1.1	0.7	2.6	2.6
20,000 and higher						
1974	5.2	6.8	1.2	0.6	1.8	3.5
1977	6.3	4.0	1.2	0.6	1.8	2.6
Homeowners						
Less than 5,000						
1974	10.4	10.1	3.5	2.4	8.1	6.7
1977	16.4	11.2	5.4	1.8	11.0	10.0
5,000–9,999						
1974	10.9	9.6	3.3	3.3	5.7	1.6
1977	13.9	12.4	4.5	2.0	7.2	5.2
10,000–14,999						
1974	10.7	12.3	3.5	6.8	4.4	4.4
1977	14.2	7.0	5.5	3.9	6.1	1.8
15,000–19,999						
1974	8.8	5.8	3.0	5.0	3.3	...
1977	13.4	2.6	7.1	2.6	4.9	2.6
20,000 and higher						
1974	10.1	8.1[e]	4.7	3.7[e]	3.4	2.0[e]
1977	12.2	6.0[e]	3.6	6.0[e]	6.2	...[e]

Source: Tabulations by the author from the users' tape of Bureau of the Census, *Annual Housing Survey*.

a. Incomes are for 1973 and 1976, in constant 1973 dollars.

b. A source of heat such as an electric space heater used in addition to the regular source.

c. A furnace breakdown causing the unit to be without heat for more than six hours more than once during the preceding winter.

d. A room closed off during the winter so that more heat would be available to the rest of the unit.

e. Based on fewer than twenty-five observations.

1.0–1.2 percent for each percent increase in energy prices.[14] Many of these estimates are based on cross-sectional data or were derived from the results of simulation models. Considerable uncertainty about the extent of the long-term response to energy prices therefore remains.

It has sometimes been argued—though there is no direct evidence—that the responsiveness of low-income households could well be greater than that of other households because of the greater share of income devoted by the lower-income households to residential energy. The price elasticity of the demand for housing by different income groups may be instructive, however, since the demand for home energy is derived from the more general demand for housing services. Follain estimated long-term price elasticities for low-income and high-income households using Annual Housing Survey data for twenty standard metropolitan statistical areas (SMSAs).[15] He found consistently *lower* price elasticities for low-income households—those whose incomes were less than four times the rent of a modest dwelling—than for high-income households among both homeowners and renters and among households headed by both nonelderly and elderly persons.

The available evidence, then, suggests moderate housing adjustments by low-income households to energy price increases. Still, changes of dwelling by renters and improvement in weatherization that is made as the owner-occupied stock turns over may lead to the higher long-term response suggested by present estimates.

Again bits of information on the type and extent of adjustments being made are available. Considering the percentage of households, by income class and age of head of household, that made various investments in weatherization in structures of one to four units in 1977, on the basis of the NIECS data, 24 percent of households whose 1977 incomes were less than $5,000 and 36 percent of all households made some improvement in weatherization in that year. The gap between households whose incomes were less than $5,000 and all households, however, yawns much greater as a percentage of the activity of low-income households—2 percent of the former and 6–7 percent of the latter—when investments in insulation or equipment classified as expensive are compared.[16] The

14. Marcuse, "Energy Pricing and Availability," table 16, p. 37; table 17, p. 37; table 18, p. 38.

15. J. R. Follain, "A Study of the Demand for Housing by Low- versus High-Income Households," paper presented at the 1979 Midyear Meeting of the American Real Estate and Urban Economics Association (San Francisco: Federal Home Loan Bank, 1979).

16. Department of Energy, *Residential Energy Consumption Survey: Conservation*, table 29B, pp. 119, 120; table 32B, pp. 131, 132.

DOE definition of expensive items includes some things, though, such as storm windows and storm doors, that are not very costly. There is little difference in this regard between households in the $5,000–$10,000 income group and higher-income households. When the age of the head of the household is considered, little difference can be seen in the incidence of the investments of households headed by persons at least sixty years old and those of other households.

It is important to recognize that the lower responsiveness of the poor and the elderly embodies more than a simple income effect. High discount rates are certainly a factor. Among the nonelderly poor, the inability to obtain financing for improvements is an impediment, and some groups—such as households headed by women—may have little idea of what should be done or how to do it cheaply themselves. The elderly as a group know less about the cost-effective alternatives, are less able to do installations themselves, and have greater hesitation in seeking outside help because of fears of being cheated and of having strangers in their homes. Many who live in units in which some energy improvements, such as the installation of storm windows, have been made consider their homes to have been weatherized.[17]

Another rough set of indicators of the extent of the adjustments that are being made is the characteristics of units removed from the housing stock in recent years. Such units have often been at the bottom of the quality distribution and have been occupied by the poor immediately before their removal. They may have been the least energy-efficient units as well. Table 9 presents figures on all the units in the housing stock in 1973 and those removed from the stock during the period 1973–78. Vacant units and rental units were removed from the inventory at higher rates than owner-occupied units. In all regions about four times as many mobile homes left the stock as were in it in 1973. While somewhat higher removal rates would be expected because of their higher rate of depreciation than that of conventional dwellings, this rate seems extraordinary. Also in the Northeast and North Central regions, more structures having two to four units and those having five or more left the inventory than there were in the base stock—perhaps reflecting in part the general problems of energy conservation in rental housing. Finally, among the units removed from the stock in all regions many more used room

17. For more information on the weatherization of homes occupied by the elderly and the perceptions of the elderly about conservation opportunities, see data from surveys in selected neighborhoods in Washington, D.C., and Denver in Wayne Hoffman, "Elderly Households and the Energy Conservation Gap," research report (Urban Institute, 1981).

Table 9. *Selected Characteristics in 1973 of Units Removed from the Housing Inventory during the Period 1973–78*
Percent

Characteristic	*Northeast*		*North Central*		*South*		*West*	
	All units (1973)	*Those removed (1973–78)*	*All units (1973)*	*Those removed (1973–78)*	*All units (1973)*	*Those removed (1973–78)*	*All units (1973)*	*Those removed (1973–78)*
Tenure								
Owned	55	21	64	34	60	40	56	36
Rented	38	65	29	45	30	41	36	50
Vacant	7	14	7	21	9	19	8	14
Units in structure								
1	55	22	71	45	76	54	67	40
2–4	21	33	13	22	7	9	11	17
5 or more	21	36	12	16	11	8	17	16
Mobile home	7	8	4	17	6	29	5	27
Heating equipment								
Central	89	84	83	68	48	35	50	40
Built-in electric	4	1	4	2	7	4	9	7
Floor, wall, or pipeless furnace	1	2	4	6	12	8	26	22
Room heater	4	8	8	20	25	38	9	21
Fireplace or stove	2	4	1	2	6	11	2	5
None	0	1	0	2	1	3	2	5

Source: Bureau of the Census, *Annual Housing Survey: 1973*, part A, *General Housing Characteristics* (GPO, 1975), table B-1, pp. 28, 33; table C-1, pp. 43, 48; table D-1, pp. 58, 63; table E-1, pp. 73, 78; ibid., 1978, part A (GPO, 1980), table B-5, pp. 86, 87; table C-5, pp. 120, 121; table D-5, pp. 154, 155; table E-5, pp. 188, 189. Percentages may not add to 100 because of rounding.

heaters, fireplaces, and stoves as their primary sources of heating than there were units remaining in the stock. Because most mobile homes are centrally heated, the incidence of room heaters and fireplaces in the conventional housing removed from the stock is even higher. While the units removed were likely to be in poor condition generally, the fact that they appear to be among the less efficiently heated may imply some reduction in the demand for energy-profligate units by lower-income households.

The Function of Public Assistance in Fostering Adjustments

The foregoing material provides a general picture of the manner and extent to which low-income households are adjusting to higher energy prices. Here the extent to which public programs are fostering residential energy conservation through improved weatherization will be considered briefly. Included are federal activities to encourage weatherization of the stock of federally assisted housing built before the era of higher energy prices, grant and loan programs for the upgrading of dwellings of which improved weatherization is not an explicit goal, the federal Weatherization Assistance Program (WAP), and tax relief focused exclusively on energy-saving improvements. With the exception of federally assisted housing, these programs have been largely concerned with owner-occupied housing. In the next section I shall examine the problems of encouraging better weatherization of the private rental stock.

Housing Assistance Programs

About 3.5 million households live in rental housing assisted by the federal government. The discussion here will be concentrated on two out of the plethora of programs: public housing and the Section 8 Existing program—as the Low-Income Rental Assistance Program is popularly known—which together accounted for about 2 million assisted units in 1980. Because the participants in these programs are heavily concentrated among households whose incomes are less than 50 percent of the median family incomes in their areas, the programs were serving about a quarter of the eligible renter households with very low incomes in 1979. About 35 percent of the participant households, moreover, are headed by elderly persons.

PUBLIC HOUSING. In 1979 there were about 1.2 million occupied units in the public housing program. These were scattered among some 2,800 local housing authorities, although about a third of the stock was concentrated in two dozen very large cities. Since two thirds of these units were constructed before 1970—93 percent before 1974— it seems likely that there would be substantial room for improvement in the conservation of energy. There is no standard arrangement between the authority and the tenant for paying for utilities; indeed it often varies from project to project within an authority. So it is difficult to make any general statements about the direct incentive to the Public Housing Authority (PHA) for conservation.

On the other hand, the method of determining the operating subsidies paid to authorities has embodied fairly strong incentives for conserving energy since 1945, when the Performance Funding System—the system for computing the operating subsidies to individual PHAs—began operation, replacing the earlier ad hoc methods of determining subsidies. The authority is reimbursed completely for increases in expenditures because of rate changes but receives only a 75 percent reimbursement for increases in consumption during a referenced period and is allowed to keep 25 percent—50 percent beginning in 1980—of savings realized from reductions in consumption. It might therefore have been expected that the PHAs would work hard to conserve energy. Those actions that require investments, however, are generally funded under a separate modernization program, allocations from which have been essentially capricious.[18]

In a study in which detailed inspections of 400 projects were made, the energy savings that could be realized through a wide range of options for the public housing stock were estimated.[19] These estimates suggest that energy expenditures could be reduced 11 percent by simply improving maintenance and operating procedures—that is, without making any capital expenditures. A reduction of another 32 percent could be realized through investments that would pay for themselves in the course of a five-year period, at an average cost of $576 per unit. If these savings had been realized in 1979, they would have reduced the energy consumption of the national stock of housing in structures of two or more units by about 5 percent.

18. Raymond J. Struyk, *A New System for Public Housing* (Urban Institute, 1980).

19. Perkins & Will and the Ehrenkrantz Group, "An Evaluation of the Physical Condition of Public Housing Stock," report to the U.S. Department of Housing and Urban Development (Washington, D.C., 1980).

These figures do not reveal the amount of improvement that has been made in recent years. Neither is there any way of knowing the extent to which the authorities are now making these investments and improvements in management. Some of the $500 million in federal funds spent annually on modernization goes to conservation of energy, as does a share of the money provided by a number of local governments out of Community Development Block Grant funds to PHAs for improvements. HUD, however, developed guidance materials only in 1981 to help the PHAs determine the possibilities for conservation, so it cannot be expected that many of them will have been realized so far. Overall, the public housing stock provides a signal opportunity for conservation, one that has certainly not been fully exploited.

There are at best two steps that HUD could take to encourage energy conservation activities besides increasing the financial support of the modernization program. First, it could share with PHAs the cost of expert consultation on opportunities for the conservation of energy. Recall that in the national study mentioned earlier significant opportunities were identified that required no investment at all. A straight fifty-fifty share program, with a maximum computed for each project, operated separately from the operating subsidy system, is an obvious administrative approach. A second step would be to permit greater concentration on investments in the conservation of energy within the current modernization program—the Comprehensive Investment Assistance Program (CIAP). The CIAP emphasizes, appropriately in most instances, the comprehensive treatment of all the rehabilitation needs, project by project. If more funds were available and the immediate returns on energy investments less impressive, this approach would be preferred, but under present conditions, more latitude to permit energy investments should be granted. A final, but more radical, step would be to make fundamental changes in the way in which operation and modernization are financed, combining the two payments and making PHAs more responsible for determining the best use of the funds available to them.[20]

SECTION 8 EXISTING. By the end of 1982 there will be about 800,000 units in this program. An eligible household obtains a certificate from the administering agency and searches for a unit at a rent no higher than the "fair market rent" (FMR), the rent established by HUD for the area as the median rent for qualifying dwellings that meet a set of minimum

20. One such system, the Fair Market Rent System, is presented in Struyk, *New System for Public Housing*, chap. 9, "Suggestions for Improvement," pp. 181–99.

physical standards. The FMR set for a unit varies according to the number of bedrooms it contains. The FMR typically cited is gross rent, but in instances in which the renter pays for some or all utilities, it is decomposed into a contract rent and a utility allowance. The physical standards have to do, for example, with sanitary facilities, space and security, structure and materials, and "thermal environment." It is important to note that the standards set for the program are minimum standards; local agencies are at liberty to go beyond them.

At the end of 1976—the time for which evaluations made for the national program are available—about 42 percent of the households receiving payments lived in single-family units, another 8 percent in row houses, 41 percent in duplexes and garden apartments, 1 percent in mobile homes, and only about 9 percent in high-rise apartments.[21] Three households out of four received utility allowances.[22]

The obvious way for HUD to influence the extent of weatherization is through the physical standards; at present these standards are silent about weatherization. They require only an adequate and dependable supply of heat to the unit and that the unit be free of serious structural defects of a sort that would permit gross infiltration of air. HUD staff members administering the program do not know how often localities have gone beyond the standards set by HUD.[23]

Very little is known about the weatherization of the multifamily housing built under the mortgage interest subsidy programs, such as section 236,

21. Margaret Drury and others, *Lower Income Housing Assistance Program (Section 8): Nationwide Evaluation of the Existing Housing Program*, Department of Housing and Urban Development, Office of Policy Development and Research (GPO, 1978), table 4, p. 33; p. 36.

22. HUD commissioned another evaluation, for which data were being gathered in the field in 1979, of the Section 8 program as it operates in urban areas. The study was not designed, however, to produce estimates of the number of participants in the existing portion of the program or of the benefits that they received; it was instead a study of a group of households who applied for the program and either did or did not become recipients of benefits during the following two months. This longitudinal panel stressed changes in the quality of housing and in expenditures for housing that came about as a result of participation. This evaluation is described in detail in James Wallace and others, "Participation and Benefits in the Urban Section 8 Program" (Cambridge, Mass.: Abt Associates, 1981).

The experience of the longitudinal households in urban areas suggests that there has been some shift to high-rise units, which constitute about 20 percent of the units newly brought into the program. A sharply smaller proportion was single-unit structures.

23. See the appendix to this paper for a suggestion for encouraging investment in weatherization.

in the latter part of the 1960s and early 1970s. The age of these projects, though, suggests poor weatherization. HUD has taken no initiative, moreover, to encourage upgrading them in this way. One consequence has been the need to add ever more subsidies to cover rising energy costs.[24] (HUD has, though, revised its minimum property standards so as to raise the energy efficiency of new projects to a reasonable level.)

Overall, it appears that HUD has paid little attention to the possibilities for conservation of energy in its programs. While concern not to impair the achievement of the main goals of the program by stressing conservation is understandable, the general lack of any real initiative is discouraging.

Unrestricted Grants and Loans

At the end of 1980 there were hundreds of programs for improvement and rehabilitation of housing sponsored at all levels of the government. At the local level many of these were spawned by the enactment in 1974 of the Community Development Block Grant (CDBG) program, which gave local governments federal funds to spend on a wide range of housing activities. In recent years about 25 percent of CDBG funds have been used for housing improvement programs. Several states are now providing home-improvement financing at below-market interest rates through the sale of tax-exempt bonds; others have direct appropriations to reduce the cost of home improvements to eligible households.

These programs differ in the depth of subsidy, the eligibility rules, the degree of quality control and consumer protection, and the stringency of the minimum, postimprovement building standards.[25] This diversity, and the lack of good record keeping, makes it impossible to make any summary judgment about the degree to which these programs are fostering weatherization.

In an attempt to address the question of the efficiency of these programs in promoting energy-conserving investments, I shall examine the amount of such investment occurring under the State of Wisconsin's

24. Figures on the energy outlays of such projects are given in John S. Williams, Jr., Dennis Eisen, and Albert Beverly, "Energy Consumption Data in HUD-Insured Multifamily Housing," report to the U.S. Department of Energy (Washington, D.C.: Metro Study Corp., 1978).

25. University of Pennsylvania, "Community Development Strategies Evaluation: Targeting to Dwelling Units," report to the U.S. Department of Housing and Urban Development (Philadelphia: University of Pennsylvania School of Social Policy, 1980).

Table 10. *Summary of Loan Programs in the Wisconsin Housing and Neighborhood Conservation Program*[a]

Terms, eligibility, and source of funds	*Installment loans*	*Deferred-payment loans*
Loan terms		
Interest rate	4–8 percent, depending on income	0 percent
Period	Maximum 15 years	Repaid when property is sold
Maximum loan	$9,000 for a single-family unit; $5,000 a unit, 2–4-unit structure	$9,000 for a single-family unit; $5,000 a unit, 2–4-unit structure
Eligibility		
Type of tenure	Homeowners	Homeowners
Type of structure	1–4 units	1–4 units
Income of household	50–100 percent of county median income	Less than 50 percent of county median income
Source of funds	Wisconsin Housing Finance Authority bonds	State appropriation

Source: Wisconsin Department of Development, *Rebuilding Wisconsin: An Evaluation of Wisconsin's Housing and Neighborhood Conservation Program* (Madison: Department of Development, 1981).
a. A somewhat more generous version of both these programs operates in locally defined "reinvestment neighborhoods."

Housing and Neighborhood Conservation Program (HNCP) between 1978 and 1980. This program was selected because there are no minimum weatherization requirements as part of the loan conditions, the program is administered by local financial institutions with a minimum of red tape, so access is quite good, and it is directed to low-income and moderate-income households. The lack of "energy standards" argues that the investments should be the lower limit to the investments made under such programs. On the other hand, the severe climate of Wisconsin means that the rate of return on energy investments will be higher than that in milder climes. On balance, the experience here should provide rough guidance on the extent to which subsidy programs are supporting energy conservation.

The HNCP consists of two loan programs for homeowners living in one- to four-unit structures. As summarized in table 10, the Installment Loan Program makes loans at 4–8 percent interest to households whose incomes are between 50 and 100 percent of the county median family income. The Deferred Loan Program, in contrast, makes interest-free loans to those whose incomes are lower than 50 percent of the county median income; the loans are repaid at the time the dwelling is sold. During the period 1978–80, 3,401 installment loans were made, the average value of which was $5,700, and 552 deferred loans were made, the average value of which was $6,600.[26]

The share of loan-financed improvements going to weatherization is substantial—about a third overall. Households headed by persons over the age of sixty devote a slightly greater share of their loans to energy-saving improvements, and there is little variation with income across the two programs. This suggests that the lack of weatherization among units occupied by low-income households is caused by low levels of investment in the upgrading of dwellings, rather than investment in improvements other than those that help to conserve energy. Low overall investment rates are borne out by national data from the Survey of Residential Repairs and Alterations. In 1979, the average homeowner whose income was less than $10,000 invested $220 in additions, alterations, or major replacements to his dwelling; households whose incomes were in the $10,000–$15,000 group spent 86 percent more; those whose incomes were more than $25,000 spent 400 percent more on the average.[27]

Weatherization Assistance

Programs focused exclusively on improving the weatherization of dwellings should differ sharply in the amount of weatherization induced from that promoted by the general purpose home-improvement programs just reviewed. Federal, state, and local versions of such programs exist. I discuss only the current federal programs—both the federal income tax credits and the provision of in-kind weatherization assistance—because better information for them is readily available. With the likelihood that the Department of Energy's Weatherization Assistance Program (WAP) will be folded into HUD's CDBG program, however, local activities are certainly of increasing interest.[28]

The WAP provides for installation of insulation, storm windows and

26. Wisconsin Department of Development, *Rebuilding Wisconsin: An Evaluation of Wisconsin's Housing and Neighborhood Conservation Program* (Madison: Department of Development, 1981).

27. Bureau of the Census, *Current Population Reports,* series P-60, no. 124, "Characteristics of the Population below the Poverty Level, 1978" (GPO, 1980).

28. For an eclectic survey of these activities, see Deborah Both and Robert Dubinsky, "Summary of Findings and Bibliography for the Department of Energy's Innovative Residential Retrofit Delivery Systems Demonstration" (Santa Monica, Calif.: Rand Corp., 1981).

doors, and other improvements in energy efficiency for housing occupied by low-income families—that is, households whose incomes are less than 125 percent of the poverty line or one of whose members received aid to families with dependent children (AFDC) or supplemental security income (SSI) during the twelve months preceding application to the program—especially the elderly and the handicapped. Both homeowners and renters are eligible, but only about 11 percent of the dwellings receiving services have been rented.[29]

The number of dwellings weatherized under the program has risen steadily since 1978: 192,000 units had been weatherized by August 1980, and the expected volume in 1980 and 1981 was 300,000 units.[30] Assuming that this goal was met, the program weatherized 838,000 units between 1978 and 1981; this added to the 400,000 done under a similar but now discontinued program of the CSA gives a total of about 1.2 million units. Figures for 1977–79 indicate that 38 percent of the dwellings assisted had elderly members.

From the materials presented earlier, the concentration of weatherization assistance on those with incomes of less than 125 percent of the poverty line is highly appropriate.[31] The cumulative effect, moreover, is substantial. Estimates by the Department of Energy (DOE) indicate that there were 5.3 million homeowners and 6.6 million renter households eligible for assistance in 1975. If 90 percent of the assistance under the CSA and DOE programs has gone to homeowners, 20 percent of the eligible owner-occupied units, but only about 3 percent of the rental structures, had been treated by the end of 1981. There was, of course, a substantial turnover and some growth during the period 1975–81 in the poverty population and changes in the units occupied by eligible

29. Department of Energy, *Residential Energy Consumption Survey: Conservation*, p. 8. Program funds can be used for the purchase of materials, administrative services, program support, and under certain circumstances, the services of laborers or contractors for installation. Through 1980, about 80 percent of the workers in the program were paid by the Comprehensive Employment and Training Act (CETA). Local weatherization projects are typically operated by a Community Action Agency, but the funds are allocated to the state by the DOE. In most cases there is a ceiling of $1,000 on the allowable expenditures per unit; this was supplemented by CETA wages. Usually about 50–60 percent of the expenditures are for materials and the remainder are for other support items.

30. Ibid., pp. 5, 43–44.

31. For a further discussion of strengths and weaknesses of the program, see Alan L. Cohen and Kevin Hollenbeck, "Energy Assistance Schemes: Review, Evaluation, and Recommendations," in Landsberg, ed., *High Energy Costs*, pp. 60–120.

households; thus, the number of households that now qualify for help may not have changed much.

The credit on federal income tax returns available for energy conservation expenditures differs from the WAP in several important features. Most obvious is that under the WAP the agency makes the weatherization improvements, while under the tax credit the household makes them itself or arranges to have them made. The WAP is tightly restricted to low-income households; the tax credits are not restricted at all, but the fact that one cannot file a short-form return and claim the credit militates against its use by low-income households. Finally, while the WAP provides a subsidy of 100 percent to cover up to $1,000 in materials, tax credits provide a subsidy of 15 percent on the first $2,000 of materials and labor. Tax credits can be used for a wider range of conservation activities than the WAP will accomplish. The programs have in common that participation is open to homeowners and renters, but in the WAP only 10 percent of the participants are renters, and a similar take-up rate for tax credits can be assumed, although no one knows.

Two points stand out from the early experience with the program, as illustrated by the figures in table 11. First, the vast majority of those taking the credits have moderate or higher incomes. Participation rates rise steadily with income; only about 1 percent of those filing returns whose incomes were less than $10,000 took credits. Second, it is a high-volume program. In the tax year 1978, which covers expenditures made during the twenty months preceding January 1979, 5.9 million households received some tax credit. Among those whose incomes were less than $10,000 almost a half million households took a credit. On the average these low-income households had spent $665 and had received a credit of $100.

One way to put federal assistance for weatherization in perspective is to compute the share of all low-income households—that is, those whose 1977 incomes were less than $10,000—who made significant weatherization expenditures in 1978 that received assistance. Assuming that 90 percent of those who make investments in weatherization and an equal percentage of those who received assistance were homeowners, I calculate that 426,000 of the 1.98 million low-income households, or 21 percent, who made investments in 1978 received federal assistance in kind or tax credits. Thus, federal assistance to homeowners has been formidable. The assistance to renter households, however, has been slight.

Table 11. *Use of Income Tax Credit for Residential Energy Conservation, by Income Class, in Tax Year 1978*[a]

Adjusted gross income class (thousands of dollars)	All returns (thousands)	Percent that made energy-conservation expenditures: Any expenditure	Expenditure for insulation	Expenditure for storm windows or doors
Less than 2	9,048	0.1	0.1	0.1
2–4	9,260	0.2	0.1	*
4–6	8,400	1.0	0.6	0.4
6–8	8,263	1.9	1.3	1.0
8–10	6,944	3.2	2.1	1.7
10–12	6,096	3.3	2.2	2.0
12–14	5,603	4.6	4.5	2.5
14–16	5,016	7.2	5.0	4.2
16–18	4,682	11.4	6.4	5.4
18–20	4,277	11.8	7.6	16.8
20–30	13,938	15.4	10.5	9.0
More than 30	8,357	18.0	11.5	10.0

Source: Unpublished data given the author by the U.S. Internal Revenue Service.
* Less than 0.05 percent.
a. Credits were permitted for the period April 30, 1977, to December 30, 1978.

Conservation in the Rental Housing Stock

About rental housing in general and multiunit rental housing most spectacularly, there is an appalling dearth of information. This is particularly unfortunate, since the poor, especially those in urban areas, are predominantly renters. I shall therefore review the basic incentives to encourage weatherization of rental housing, then look more closely at the situation of properties occupied by the poor.

The Basics

The amount of energy consumed in a rental structure depends on both the energy efficiency of the structure and mechanical systems and the energy-consumption practices of the residents.[32] The landlord owns

32. For this section I have drawn heavily on Alisa Gravitz, "DOE Report on the Rental and Multi-Family Housing Energy Efficiency Workshop," memorandum, U.S. Department of Energy, Office of Conservation Policy and Evaluation, 1981.

the property and controls that part of the consumption of energy that is related to the structure and mechanical systems of the building. On the other hand, tenants live in the building, and control the amount of energy consumed in their respective units. Thus, the responsibility for total building energy consumption and control of it is divided between landlords and tenants. As a result, when energy prices rise, no mechanism exists for trade-offs between property improvements and changes in habits of consumption in the effort to save energy.

The arrangements landlords and tenants have developed for paying energy costs also impede their response to increasing energy prices. There are three principal arrangements between landlords and tenants for payment of energy bills:

1. *Tenants pay all the energy bills* directly to the utility companies.

2. *Landlords pay all the energy costs.* In some instances landlords can pass along all energy price increases, depending on lease agreements, rent controls, and market conditions.

3. *Tenants pay some of the energy costs, the landlords the rest.* Typically, electricity is individually metered to the tenants, while landlords pay the oil and gas bill for central heating systems. Sometimes gas for water heating and cooking is also billed directly to tenants. Here again, landlords can sometimes pass on energy costs through the rent.

A good idea of current arrangements for paying utility bills is available from the AHS data for 1977 shown in table 12, which shows the percentage of renters who pay their own electricity, heating fuel, and both electricity and heating fuel costs directly to the utility companies. A somewhat lower share of low-income households—those whose income was less than $5,000 in 1977—than of higher-income households pay these bills directly. Direct payment of utilities is much less common in multifamily units, especially those in structures having twenty units or more, than in single-unit structures, although even for 17 percent of these units heat is included in the rent. Among the large multifamily properties, households seldom pay their heating fuel bills, and the incidence is especially low for households whose incomes are very low. Thus, while there is negligible variance in the distribution of renters among structure types with income, the renters whose incomes are lowest more often pay for their utilities as part of their gross rent.

If landlords can protect themselves from having to pay energy costs, they have little incentive to improve the energy efficiency of their buildings. A review of the arrangements listed above shows that landlords

Table 12. *Percentage of Renter Households Who Pay Their Own Energy Bills, by Income Class and Number of Units in Structure, 1977*
Income classes in dollars

Income class and number of units in structure	Electricity	Heat	Electricity and heat
Income class			
Less than 5,000	69	51	50
5,000–9,999	77	57	56
10,000–14,999	80	60	59
15,000–20,000	82	59	59
More than 20,000	84	59	58
Number of units in structure			
Single unit	89	83	82
2–4	81	60	60
5–19	72	43	43
20 or more	58	23	23
Number of units in structure, by income class			
5–19 units in structure			
Less than 5,000	61	36	35
5,000–9,999	73	45	44
10,000–14,999	76	47	46
15,000–20,000	77	45	45
More than 20,000	81	47	47
20 or more units in structure			
Less than 5,000	44	15	14
5,000–9,999	59	26	26
10,000–14,999	62	27	27
15,000–20,000	69	30	30
More than 20,000	70	25	25

Source: Author's tabulations from Annual Housing Survey data tapes.

can completely shield themselves from the effects of rising energy prices when tenants pay all the energy bills or when landlords can pass on all energy costs through the rent. In buildings in which these arrangements prevail, price pressures will not directly motivate landlords to make improvements for the sake of energy efficiency.

In these instances, owners perceive little opportunity to earn returns on their energy investments through cost savings. In order to earn a return, an energy investment would have to allow the landlord to charge a higher rent net of utilities for energy-efficient units. In concept, this should be possible. With the price of a bundle of housing services established by the market—whether utilities are paid partially or completely to the landlord—more efficient landlords should be able to charge the same prices as their less efficient counterparts and realize positive returns from energy-conservation investments.

Through a longer time span inefficient landlords who are able to pass energy costs along to the tenants may receive additional signals from the market. Depending on the magnitude of the price elasticity of demand for rental housing services, tenants will shift away from those buildings that cost more per unit. Reduced demand should induce the landlord to make the investments that would reduce his price per unit of service or to withdraw the unit from stock. The information available from the Demand Experiment of the Experimental Housing Allowance Program on the price elasticity of demand indicates an inelastic demand, at least among lower-income households.[33] Even those landlords who pay the energy bills and should be motivated to make improvements for the sake of energy efficiency, however, often face severe cash-flow problems and have difficulty obtaining capital for efficiency improvements.

When tenants pay their own energy bills, they feel the pressure of rising prices. They can control their consumption of energy, however, only by changing their way of living. Renters are unlikely to improve the energy efficiency of the building, because any investments they made would become part of the landlord's property. Many investments would require the permission of the landlord, and organizing fellow tenants to participate in demanding and possibly installing the improvements would require a good deal of effort by a few tenants. Many important energy-efficiency measures, moreover, such as improvements in the control of heating and air-conditioning systems, can only be taken by the landlord. The estimates of short-run price elasticities, therefore, are probably a reasonable indicator of the extent of adjustment that can be anticipated for structures in which renters pay their own utility bills.

On the other hand, when rent includes energy costs, tenants cannot see the direct effects of their actions on energy consumption, especially since rents are usually raised infrequently, and energy costs are spread among all tenants. As a result, energy price increases do not even inspire changes in energy-consumption practices, as they do when renters pay energy bills directly.[34]

33. See Francis J. Cronin, "Household Responsiveness to Unconstrained Housing Allowances," in Raymond J. Struyk and Marc Bendick, Jr., eds., *Housing Vouchers for the Poor: Lessons from a National Experiment* (Urban Institute, 1981), pp. 159–76.

34. A suggested demonstration for sharing savings in energy costs between owners and tenants in centrally metered properties is sketched in the appendix to this paper.

Other Factors That Affect Adjustments

While the preceding section provides a general picture of the situation in rental housing, other factors affect the degree to which landlords who provide housing to low-income households respond to price signals. First, many of the rental properties occupied by low-income households in urban areas are in marginal areas—areas characterized by little if any appreciation in property values, which generally discourages owners from investing in their structures. Regardless of location, it is likely that the owners of these properties are suffering acute cash-flow problems. Lowry, for example, offers a description of the deteriorating position of rental housing during the 1970s.[35] This means that there are few funds already within the project that can be used for financing investment. Financial institutions are understandably hesitant to provide financing and often demand refinancing of the outstanding debt at current interest rates as a condition of the provision of financing for the improvements.

Another impediment is the rate of return on the investment. For single-family and two- to four-unit properties, there appears to be sufficient experience that savings to be expected from conservation investments can be calculated with some confidence, although even here technical advice is probably essential for being certain that the most cost-effective expenditures are made first. The likelihood that owners of a few units will seek such advice seems scant. For larger structures, the degree of uncertainty of the savings from particular improvements is much greater. Indeed, the lack of experience (as opposed to simulated results) is often cited as a primary factor inhibiting investment in the larger properties.[36]

Realization of an adequate rate of return on energy-conserving investments and the ability to borrow funds are both made more uncertain and difficult by rent controls. This problem exists disproportionately in the Northeast, where many of the profitable investment opportunities are concentrated.

Regardless of the size of the property, the owners require competitive rates of return and are looking for repayment periods of no more than one to three years. In calculating those repayments it must be borne in

35. Ira S. Lowry, "Rental Housing in the 1970s: Searching for the Crisis," draft, WN-820-1-HUD (Santa Monica: Rand Corp., 1981).

36. See, for example, Real Estate Research Corporation, "Energy Conservation and Existing City Buildings: Owners and Retro-fit Investment," report to the Office of Technology Assessment (Washington, D.C., 1981).

mind that higher income will be subject to taxes—offset to some degree, of course, by deductions for depreciation—while higher energy costs shelter incomes.

As suggested earlier, the demand-side pressure for adjustments may be less than might be expected. The low degree of renters' responsiveness to the price of housing services per unit reflects the lack of "perfect information"—a prerequisite for a competitive market—and there is at least fragmentary evidence of segmentation of the market. In properties in which households pay their own utilities, moreover, it may be quite difficult to obtain full information on gross rents. Low-income households are known to rely heavily on friends and relatives for finding units when they move, and many find units in this way without really searching.[37] This is another reason that competition is weak and information is difficult to obtain.

Another factor that may affect both the rate of investment and the overall conservation strategy is the relation between the owner and the tenants. Relationships are presumably closer in properties that contain only a few units, one of which is occupied by the owner or the owner of which lives in the immediate neighborhood.[38] As property size grows and contact with the owner declines, the quality of owner-tenant relations can be expected to decline. Where the relation is closer, the trade-offs between investments and behavioral changes are more likely to be worked out jointly and a decision process closer to that of the homeowner approximated. Interestingly, the incidence of owner residence in one unit of structures that have two to four units varies little with the gross rent of the unit; hence, low-income renters are as likely as other households to occupy units in owner-resident properties.

On balance, motivating owners of rental housing that is occupied by low-income households to make energy-conservation investments is only moderately different from motivating them to make other investments. Even where the tenants pay for their utilities directly, competition should push owners to make these investments if they offer a higher rate of return than other opportunities in the property or in the market at large. In determining how best to encourage investment, it should be possible

37. Kevin F. McCarthy, *Housing Search and Mobility*, Rand Report R2451-HUD (Santa Monica: Rand Corp., 1978).

38. Peterson, using St. Louis data, for example, after controlling for other factors found that rents were significantly lower in properties in which the owner lived. George Peterson, "Housing Prices and Tenant Characteristics: Analysis of St. Louis Housing Data" (Urban Institute, n.d. [1972?]).

to learn a good deal from the effectiveness of various tax credits and deductions and from grants and subsidized loan programs, such as S.312 and CDBG-funded programs, in motivating investment generally in low-income properties. Remarkably no comprehensive review of this type has to my knowledge been undertaken; indeed, the effectiveness of the individual programs has seldom been examined in any real depth.

A Possible Future

In this section I shall describe fragments of the possible adjustments in the stock of private housing existing in 1980 that may take place in response to higher energy prices. The projection is not a set of predictions. Rather it is a set of possibilities presented to stimulate discussion and debate. The attempt here is to go beyond the multiplication of price increases and estimates of the long-run price elasticity of demand for residential energy in describing the response to price increases. The focus is on the adjustments made by homeowners, landlords, and renters in various types of structures that are expected on the basis of the foregoing discussion. I shall concentrate on the existing stock because it is this stock that low-income households will occupy.

My time projection is quite long—perhaps ten to fifteen years. I shall assume that the price of energy will continue to rise faster than the composite price of other goods and services but more moderately than the explosion during the latter half of the 1970s. Stated another way, my assumption is that significant price pressure for housing adjustments in response to higher energy prices will remain. The other general assumption is that the rate of new construction in housing will return to a steady state 1.8–2.2 million units a year in about five years. The occupancy cost of housing will be higher in real terms, thanks to 1980 legislation that reformed the terms on which thrift institutions compete for funds. High-income renters will be devoting a greater share of their income to housing—say 20–22 percent—and higher levels of privately sponsored development of new multifamily housing will be under way.

As a standard for weatherization I shall use those few investments that offer very high returns under present conditions. It is quite possible that the stock may be improved to this level during the next decade but that by then significantly more weatherization will be justified.

The approach taken here is to discuss the effects of adjustments on

Table 13. *Number of Households Living in Selected Types of Structure, by Type of Tenure, Income Class, and Age of Head of Household, 1979*
Thousands

Type of tenure, income class, age of head of household, and size of structure	*Total*	*Location: Inside SMSAs: Central city*	*Location: Inside SMSAs: Other*	*Location: Outside SMSAs*
Homeowners				
Elderly head of household	11,607	2,847	3,957	4,803
Low-income nonelderly head of household	2,330	865	742	723
Renters				
In single-family units				
Household in poverty[a]	2,639	768	632	1,239
Others	6,520	1,430	2,858	2,232
In structures with 2–4 units				
Household in poverty	2,813	1,467	639	707
Others	6,131	2,152	3,061	918
In larger rental structures				
Household in poverty	3,661	2,363	829	469
Others	9,326	3,689	5,091	546
All occupied housing units	78,572	23,697	29,944	24,931

Source: *Annual Housing Survey: 1979*, part C, *Financial Characteristics of the Housing Inventory for the United States and Regions* (GPO, 1980), table A-1, pp. 1–35.
a. A family income of $7,000 is used as the poverty cutoff.

five groups of households and the units they occupy at this writing: elderly homeowners, low-income nonelderly homeowners, low-income renters in single-family units, renters in two- to four-unit properties, and renters in larger properties. Table 13 provides counts of households in selected types of tenure and structure to give an idea of the number of households being considered. The choice leaves some formidable gaps, but it should be sufficient to give a general idea of what the future may hold.

Elderly Homeowners

The story here is a familiar one. Elderly couples and individuals continue to live in their single-family homes long after their families have been raised. Three out of four elderly-headed households are homeowners and five out of six of these have no mortgage debt. They have remained because of the familiarity of their neighborhoods and the control they have been able to maintain over their out-of-pocket housing costs. Energy price increases have, however, eroded this control in recent years and have increased the share of their resources that must be devoted to housing. As noted earlier, low-income elderly are similar to other low-income households in their energy expenditures and in their conservation activities. Typically, the poor spend little to improve their housing. In the absence of assistance with weatherization expenditures not much will be done.

At the same time other developments affect their choice of housing. The cost of relocating to a rental unit or cheaper owned unit has been reduced by the enactment in 1978 of legislation granting forgiveness of the first $100,000 in capital gains on the sale of a home by people over the age of fifty-five. The widespread availability of condominiums offers another situation of "controllable" costs, but with a smaller unit. On the other hand, reverse annuity mortgages, if they gain acceptance, will provide cash resources to permit many of the elderly to remain in their homes.

On balance, the elderly would be expected to begin shifting out of their single-family units at significantly higher rates. This will be fostered by a continuing demand for single-family units as the children born during the baby boom age into parenthood. The new owners, with greater propensities to invest, will make the economically rational weatherization improvements neglected by the low-income elderly. From the figures presented earlier the potential energy savings from this one-time shift and upgrading seem to be moderate. By the end of the period the process will be virtually complete, but the earlier relocation by the elderly suggested here will mean that the majority of it will have occurred during the first decade. The pattern just sketched means that the next generation of elderly households will live in better-weatherized structures and, holding energy prices fixed in real terms, will have to make lower expenditures for energy. This, in time, will lessen the pressure to relocate from single-family units.

Low-Income Nonelderly Homeowners

Perhaps the principal observation to be made here is that this group will constitute a declining share of all homeowners. The cost of becoming a homeowner has risen rapidly during the past several years as anticipated capital gains have been capitalized into sales prices and the inflation premium has been built into mortgage interest rates. Even for those lower-income households willing to devote an extraordinary share of their income to housing, homeownership may prove elusive. While a reduction of the rate of increase in the costs of homeownership can definitely be expected, these costs will continue to rise at a rate at least as great as the rate of inflation.

In general, there may be less pressure on present low-income nonelderly homeowners to sell their homes than on their elderly counterparts, so the turnover process will take longer. In urban areas at the time of sale, units in decent condition, favorable locations, or both will go to higher-income households who will improve their energy efficiency. Those in less favorable circumstances will be sold at low prices or be abandoned. Hence only a modest share of these units will remain in the active housing stock without improved weatherization beyond the end of the period.

In rural areas low-income homeownership will persist at higher levels. Assistance with weatherization, if it is continued and well directed, could have a formidable effect on this group. Otherwise, these households, unless their propensity to weatherize increases, will be particularly burdened by energy costs.

Renters in Single-Family Homes

This is another group expected to decline in importance in the future, and again the results will be better weatherization. With the increased returns on homeownership, there is general pressure for single-family rental housing to be converted to homeownership. Indeed, the process is clearly under way. The number of single-family rented units declined by 0.5 million units, or 7 percent, from 1970 to 1978, while the stock of housing rose 22 percent.

In single-family rental status, the occupant typically pays the utility bills directly. Thus there is an incentive for the resident to conserve energy but not to make investments in weatherization improvements. In some instances there may be a good relationship between the tenant and owner through which the tenant is able to realize some improvement or be reimbursed for the cost of materials if he does the work. But these seem to be atypical arrangements, although that is not certain.[39] In any event, it is expected that single-family rental units will continue to be converted to homeownership, particularly in urban areas, at which point weatherization will be improved if it has not been previously. Among those properties that remain in a rental status basic weatherization improvements might be expected to have been made, especially in those units that bring higher rents, because the occupants would demand help with their energy costs and because the landlord could afford to provide the help.

Renters in Properties Having Two to Four Units

The three preceding sections have a common theme: low-income and moderate-income households, both elderly and nonelderly, will occupy single-family housing less frequently in the future, either as homeowners or as renters. Most of these households will be found renting units in small or large properties, while single-family housing becomes more of a middle-class fiefdom. This shift to rental status will generally imply a shift to smaller units that use less energy and will bring about a one-time reduction in energy outlays for the families involved. The question then becomes whether other adjustments will be made to offset rising energy prices and solidify these gains. Of course, renters already in these structures will not enjoy this one-time "improvement."

It is typical for tenants in structures having two to four units to pay their own energy bills, with the incentives already set forth. An important

39. Information is available from the Housing Allowance Supply Experiment on who made the repairs required to qualify rental units of all types that initially failed the physical standards of the program. James McDowell, in *Housing Repair and Improvement in Response to a Housing Allowance Program,* Rand Paper P-6076 (Santa Monica: Rand Corp., 1978), gives the following analysis (in percent):

	Brown County (Green Bay)	*St. Joseph County (South Bend)*
Who made the repairs?		
Landlord	35	40
Contractor	8	10
Tenant	47	38
Friend	10	12
Who paid the bill?		
No cash laid out	34	33
Tenant	21	28
Landlord	43	37
Other	2	2

distinction for the extent of weatherization of the unit is whether the owner occupies one of the units. If he does, the union of interest that exists among homeowners may be approximated between landlord and tenant in effecting energy savings. The owner will make improvements that benefit all occupants—attic insulation for example—and each of the occupants has strong incentives to conserve his own energy use. Also, it seems that this type of landlord is less sensitive to neighborhood conditions in making investments, possibly because careful screening of tenants allows him to reduce his maintenance costs per unit below his nonresident-owner competitors and have a higher rate of return on the same rent roll.

For the absentee owner, cash flow and neighborhood conditions are more important determinants of the investment decision. Even where the tenants are paying the energy bills, if neighborhood conditions seem to be favorable for appreciation and the rent roll is strong, weatherization improvements may well be made—in part because the owner knows, or thinks he knows, what to do and in part because he thinks he can recover his outlay if he sells the property. Under less favorable market conditions, little investment can be expected. Eventually, of course, these properties become the marginal ones, having the highest gross price per unit of housing services. How long will they remain in the active stock? The answer probably depends more on the rate at which the area in which the unit is located changes its desirability than on the price competition per se. As noted, the price elasticity of demand for housing by low-income households is only about -0.5; consistent with this is the fact that many low-income renters have been observed to search for units essentially in their current neighborhood. Eventually, the less efficient units should be pushed out of the stock, but the timing depends on the rate of deterioration of the neighborhood, which itself depends on the availability of other housing opportunities in the market. In any event, it can be anticipated that the energy-inefficient rental units in two- to four-unit and larger structures now occupied by the poor will be phased out of the market; the incidence will be higher where the housing markets are generally soft. Thus particular problems will arise in the cities in the Northeast and North Central regions.

Larger Properties

In properties having five or more units, it is the rule that heat is centrally provided, and in properties having twenty or more units electricity is typically metered centrally as well. Central metering is more common in properties with low-income tenants. In this arrangement there is a strong incentive for the owner to make energy-saving investments, but there is little incentive for the tenant to restrain his consumption.

The small owner, individual or partnership, that owns a few properties in this size category, but none that contains more, say, than twenty-five units, is not likely to undertake much in the way of improving energy efficiency. Cost-effective investments in larger structures are less obvious than in smaller properties, and the relatively small number of units in these properties can make the cost of expert consultation per unit high. The small investor may also have difficulty financing these improvements, and accumulating the funds out of the cash flow may take more discipline or bring a lower return than at least some of the investors are willing to tolerate. The one item in which investment may be made is individual meters, especially in those properties where they can be installed at modest cost; this will reduce energy consumption by altering the tenants' behavior but will not alter the efficiency of the structure. Such properties are likely to drift along until they are sold and one of three dispositions is made: conversion to condominium ownership; sale to other investors, possibly at a large discount because of the unfavorable rate of return compared to that on more efficient units; or abandonment. Which is made, of course, depends on local market conditions. Greater energy efficiency will be the result of conversion or of sale to a more sophisticated and better-capitalized investor. Sale to another small investor who makes no improvements will probably set the property on the road to withdrawal from the stock; the length of the road depends on the same factors discussed in relation to the smaller properties.

The more sophisticated owner—a corporation or partnership— is more likely to make energy-conserving investments. The timing of these investments, however, will be concentrated at the time of acquisition or refinancing—points at which additional equity and credit to finance the improvements can be raised at a cost that is lower than incremental financing. Owners of these properties will have more units over which to amortize expert advice and will be sophisticated enough to seek it and to concentrate their investment on the items that will bring the highest returns. This cheering outcome will materialize, of course, only if a reasonable rate of return is in prospect. If such a return is not likely, these owners can be expected to sell the properties—at big discounts if need be—and take the tax losses.

Rental Housing and Federal Housing Assistance

A third of the renters whose incomes are below 125 percent of poverty-line income live in federally assisted housing, so the weatherization of this stock will certainly be a significant determinant of the home energy use of the poor. The "energy burden" of these households, though, will generally be unaffected, because participants in the program pay a fixed share of their incomes—generally 30 percent—as rent and the government pays the rest. There is no way to predict the action that will be taken to improve the weatherization of this now inefficient stock. There has been little movement so far, reflecting an opportunity amazingly missed.

Summary

During the next ten or fifteen years the existing stock of single-family housing will be weatherized to varying degrees; most of the high-return investments will presumably have been made by then. The outcome for two- to four-unit and multifamily stock, most of which is rental, is more complex. The primary factor is the expected rate of return anticipated by the owner on his total housing investment. Under identical market conditions, some owners realize a higher rate of return through greater diligence in selecting tenants and controlling their behavior. Where prospects for returns are good, energy-efficiency investments are likely to be made—particularly by sophisticated investors and in centrally metered units. Individual metering may be found where returns are expected to be adequate but less favorable. Under even less favorable conditions, little weatherization will be done. Given the patterns of shopping for housing by the poor, though, it can be anticipated that these units will remain in the active stock for years. This suggests a continuing need for energy-assistance payments for renters in energy-inefficient structures, although the cost of prolonged "excess consumption" must be recognized.

How do low-income and elderly households fare? Elderly homeowners will do better than the others among these households, since as they shift from larger to smaller units—often well-weatherized apartments, sometimes in condominiums—their residential energy consumption will decline. As for the others, they will divide into two groups: those who find their way into energy-efficient units and those who do not. The former group will be characterized by those with good records as tenants—or prospects of being desirable tenants—whose incomes are somewhat higher but still low and those who spend more effort searching for housing or begin their searches with an unusual array of good opportunities. In short, those whom housing economists have traditionally regarded as disadvantaged in the housing market will be those living in the least energy-efficient dwellings and paying a large share of their incomes for the privilege.

Appendix: Suggestions for Encouraging Weatherization in Assisted Housing

As noted in the text, making the standards of the program more stringent with regard to weatherization is the obvious way to proceed. For units in structures that have one to four units it would be fairly easy to go somewhat further than the present standards by requiring that steps be taken to tighten the structural envelope—weather stripping, caulking, attic insulation, and storm windows—varying the requirements with the severity of the climate. The FMRs and utility allowances could be set with reference to units having this equipment, moreover, thereby reducing the subsidy payments for utilities. Formulating the guidance for larger structures would be more difficult, as the most cost-effective improvements may well have to do with boiler efficiency and the like and not with the envelope. Still, since 80 percent of the units in the program are in structures that have one to four units, much could be done without addressing the larger properties.

The improvement envisioned here could, however, be purchased at the cost of more difficult program administration that would arise from the unwillingness of the landlords to participate. Data from the Housing Allowance Supply Experiment show that landlord participation falls off sharply as the number of deficiencies found by an inspection increases; about 70 percent of the units that fail one item eventually qualify, but only 30 percent of the units that fail four or more do.[40] A lower rate of landlord participation will increase household search times and could produce adverse effects on household participation. These effects could, though, be studied effectively through a modest experiment.

A second possibility for HUD would be to offer less than full compensation for energy expenditures in the utilities allowance delib-

40. James P. Zais, "Repairs and Maintenance on the Units Occupied by Allowance Recipients," in Struyk and Bendick, *Housing Vouchers*, table 8.5, p. 192.

erately. This would create a modest shopping incentive, push the occupants to conserve energy, and perhaps press owners of inefficient properties to make some investments. In 1976, in fact, the average utility allowance paid $7.50 a month less than actual expenditures, and less complete data for 1979 indicate that this pattern still exists.[41] A problem central to this approach is the unequal treatment of participants depending on whether they pay for any of their utilities directly.

A difficult problem is to motivate tenants in centrally metered projects to change their habits and use less energy. One obvious way of doing this is to share any savings in expenditures for utilities with the tenants, perhaps on a fifty-fifty basis. The tenants would be notified of the program through an announcement that would give examples of the size of the payments they might receive; Seligman and his colleagues suggest that setting a difficult goal as well would be helpful.[42] Thereafter tenants would receive checks periodically for their shares of the savings, along with an explanation of the computations; more frequent feedback, perhaps bimonthly, would be preferred. Savings would be calculated for each utility as the difference in the quantity of energy consumed—kwh for example—in the calendar quarter of the base year and that consumed in the current quarter times the current price per unit.

This plan presents two potential problems. First, there is no direct link between an individual household's energy consumption and its payment. This clearly reduces the incentives but on the other hand may generate social pressure against the most blatant wastage—which could by itself generate significant savings.

The second problem is to control for variations in climate in calculating savings. It seems probable that a simple algorithm could be developed to control for variation in year-to-year weather severity. Perhaps a linear relation between Btu and degree days for various types of structure would serve.

While this approach has not been widely taken, it certainly deserves extensive experimentation as an alternative to metering individual units.[43]

41. Drury and others, *Lower Income Housing Assistance Program*, p. 37; Wallace and others, "Participation and Benefits," p. 306; these data are only for new participants who came into the program in urban areas in 1976.

42. Clive Seligman, John M. Darley, and Lawrence J. Becker, "Behavioral Approaches to Residential Energy Conservation," in Robert H. Socolow, ed., *Saving Energy in the Home: Princeton's Experiments at Twin Rivers* (Ballinger, 1978), pp. 231–62.

43. The Department of Energy has planned demonstrations that go beyond the one proposed here in that they involve owner-tenant contracts for sharing in the costs of investments in the conservation of energy as well as savings from any source.

Metering has several unattractive features: it is expensive to install—Ferry suggests $100 to $1,000 per unit for full metering[44]—and the money expended could be invested in other energy-saving capital investments. As has been shown, it also reduces the incentive to owners to make energy-conserving investments, and the switch often causes rates to be higher as bulk discounts disappear.

Comment by Kenneth J. Saulter

This paper surveys the available statistical data on energy expenditures and conservation behavior of households classified as poor or elderly. The general conclusions are consistent with a large body of statistical data and reports on this subject. A few of the principal findings include:

Concerning the poor

—Poor households spend a much higher share of their income on home energy than do other households. This regressivity appears to increase in magnitude for the lowest income levels.

—The dwellings of poor households are less well weatherized, although weatherization is generally widespread among low-income households in the coldest regions of the United States.

—A higher share of low-income renters in multiple-unit buildings do not pay for their own utilities.

—The poor do not seem to differ from the nonpoor in price responsiveness.

Concerning the elderly

—Household units occupied by the elderly do not differ from the average in weatherization.

—There is no real difference between elderly and nonelderly households of the same income levels in the share of income spent for energy.

—About half of all households that spend a large share of their income on energy are elderly.

The author believes that low-income households will increase the rate at which they shift from homeowner status to renter status and will continue to shift from single-unit housing to multiple-unit housing. In addition, he believes that it is not so much the elderly per se who present

44. Steven Ferrey, "Fostering Equity in Urban Conservation: Utility Metering and Utility Financing," report to the Office of Technology Assessment (Washington, D.C., 1979).

a housing issue as it is the narrower group of elderly poor renters, especially of energy-inefficient units.

The central difficulty about which the author appears to be concerned is the landlord-tenant incentive relationship with respect to investment in weatherization. Implicitly he regards this as both a perplexing problem and a growing one, in view of the expectation that more and more of the poor and the elderly will be renters and will live in multiple-unit dwellings in which energy use may not be individually metered and incentives to investment in weatherization are insufficient to promote such investments on a large scale.

The text is detailed and laden with statistics. The central hypothesis is that the share of income spent for energy by all poor households will increase substantially in the future. The author's central task is to trace some implications of this hypothesis, for the most part qualitative, for the type of housing the poor will occupy by 1990.

The problem with survey studies, heavily laden with statistics, is generally a lack of focus and a lack of context. The present study is a mural where, in my view, a portrait would be more appropriate. I say this having written my own mural studies and having become convinced that, for policymaking or practical purposes, it may be much more useful to focus attention on one or two aspects of the problem and to balance the statistical picture with more institutional information. This has been done in the present paper for the case study of the Wisconsin weatherization-assistance program. Two such problems that deserve closer attention, for example, are low-income fuel oil buyers, especially in the Northeast, and low-income rural households in the Midwest and the Plains states. These, in my view, are two particularly vulnerable groups of energy consumers worth studying. But attention should be given not only to the statistics but also to institutions, in that program needs or program deficiencies with respect to these two groups should be identified and evaluated.

In short, the present study is difficult to evaluate because no thesis is offered. Several problematic issues are touched upon, but a lack of focus is evident. In addition, the subject matter requires a common-sense context. Consider the following generally well-known facts that are directly related to the subject of this paper but are not mentioned:

—The Reagan administration de-emphasizes regulatory or bureaucratic interventions where energy allocation is concerned and emphasizes decontrol of prices and allocation of supplies in the energy market. To paraphrase various policy statements by administration officials, moreover, it is the view of the Reagan administration that energy price controls and regulations were the energy problem of the past. Under such a regime, the energy tail wags the economy dog. With decontrol it is the economy dog that wags the energy tail. In short, the administration's emphasis is on overall economic policy, which, appropriately, will completely dominate any specific domestic energy policy program—including energy assistance to the poor.

—The Natural Gas Policy Act of 1978 will decontrol natural gas wellhead prices by 1985 if new legislation to accelerate decontrol has not been enacted before that time. Note that during the 1970s fuel oil price increases were the most pressing problem for the poor throughout the country. In the near future it could be natural gas price increases. Finally, it is likely that the wide regional disparity in natural gas prices—highest in the Northeast, lowest in the West—may disappear, as it has in fuel oil prices.

—There is virtual depression in the housing market, with new housing starts, at about 1.1 million units, possibly at their lowest level since the Depression. The only years since 1953 in which new housing starts fell below 1.2 million units were 1957 (1.173 million units), 1966 (1.167 million units), and 1975 (1.160 million units).

—High interest rates and home financing costs have prompted a proposed revision of the consumer price index and, along with banking regulatory reform, have created substantial institutional changes during the last five years in financial, mortgage, and real estate markets.

—There is a population shift to the Sun Belt that is associated, in part, with the growth of energy mining and manufacturing industries. In contrast are the areas of the industrial Midwest where the recession is deep and unemployment is high and where automobile production and its supplier industries have been traumatized by two years of financial and competitive losses.

I stop here, recognizing that some if not all of these conditions that surround the present-day housing market can be accounted for quantitatively in assessing their effects on housing for the poor. Nevertheless, in my view they are significant enough to be discussed qualitatively.

Technical Issues

These have to do with interpretations of energy demand. In particular, demand elasticities for natural gas and electricity are regarded as short-term elasticities. If these are derived from econometric estimates from cross-sectional data they should be interpreted as long-run elasticities.

Since the early consumption studies of Houthakker and Taylor it has been recognized that cross-sectional data, even covering only one or two years, represent a spectrum of production technologies, or of consumer preferences, that each has evolved during long periods.[45] Thus, a snapshot look at such a variation in technology or preferences represents a view of long-term developments—a mixture of the newest and oldest technologies or of the lowest and highest income groups. Cross-sectional estimates, therefore, have traditionally been treated as structural long-term parametric estimates.

With time-series data the situation is structurally different and the issue of long-term trend versus short-run cycle or the issue of lagged behavior enters the argument. Appropriate statistical techniques, many of which were introduced by Houthakker and Taylor, have been developed to deal with these short-run versus long-run parametric characteristics.

Thus, the fact that a cross-sectional sample represents only one or a few years does not imply that the resulting parametric estimates are short-run elasticities. This is a confusion of time-series and cross-section interpretation.

—Throughout the discussion of elasticities no mention is made of income elasticity, which should be taken into account in speculation about expenditure shares. Statistics for 1978–80, for example, indicate that federal transfer payments increased 32 percent while total personal income increased 25 percent. Given that the CPI index also increased 25 percent, the real purchasing value of transfer payments increased about 7 percent between 1978 and 1980. The point, however, is that it is not appropriate to ignore possible increases in the income of poor and elderly households during the next decade. If their income increases, there will, in general, be a nonzero income elasticity of demand for home energy that should be taken into account.

—This leads to the general issue of income and substitution effects of energy price increases. If, as supposed, energy demand is price inelastic, then as prices increase the expenditure share increases proportionately. This implies that some other expenditure share declines as a result of the so-called real income effect of an energy price increase. The question

45. See Hendryk S. Houthakker and Lester D. Taylor, *Consumer Demand in the United States: Analysis and Projections*, rev. ed. (Harvard University Press, 1970).

is, What is the magnitude of this real income effect and price effect on total consumption? That is, for a given energy price increase, by what magnitude have real standards of living of the poor declined in relation to those of the nonpoor?

Increases in food prices—2.54 percentage points in 1980 and an average of 2.4 percentage points between 1973 and 1980—have been a much more serious problem for the poor. In addition, after food, rent, fuels, and medical care have been accounted for, most of the inflation burden of the lowest-income household has been in "other" categories of consumption, which include furniture, apparel, public transportation, entertainment, and other goods and services. Prices in this category increased 4.39 percentage points in 1980 and an average of 3.03 percentage points between 1973 and 1980.[46]

Underlying these technical issues is the lack of a conceptual or theoretical model of housing demand by the poor and the elderly—a model in which the effect of higher energy prices could be analyzed, even if they cannot be measured. This is necessary if for no other reason than to check our logic and to reveal in clear terms how energy costs will affect housing costs and housing services to low-income households.

Theoretical and Empirical Issues

It is striking that no information is presented on the housing-cost share of the income of low-income households and how energy costs are related to housing costs. For example, the latest comprehensive consumer survey data for 1972–73 indicate that the average urban family allocated its expenditures in roughly the following manner:

	Percent
Energy expenditure	7
Food expenditure	20
Housing expenditure	29
Transportation, not including fuel	12
Other items	32

These data imply that average urban families spent 35–40 percent of their disposable income for housing and energy. Similar data concerning

46. Data Resources, Inc., simulation of components of the consumer price index, 1973–80, for consumers whose incomes were less than $3,000 in 1972.

low-income families are not readily available or comparable, since the average poverty-level household apparently consumes about twice what it receives in earned income. The difference between earned income and consumption is attributed to noncash transfers, unemployment payments, and other sources of funds such as savings or loans. Nevertheless, given that energy purchases are typically estimated to account for 20 percent or more of the disposable income of low-income households, total spending for housing and energy may easily exceed 50 percent of disposable income.

At this point it becomes important to consider energy inflation as a component of overall inflation and its relative effects on the poor. Among some statistics that indicate relative inflation rates for the period 1974–80 for the lowest income group in the United States and for all urban consumers, the following points stand out:[47]

—For the poor, food price inflation has been the most significant source of increases in the cost of living during this period.

—For the poor, the energy share of overall inflation has averaged 20 percent a year, with a range from 12 percent to 33 percent. For all consumers the share of overall inflation attributable to energy has averaged 14 percent, with a range from 8 percent to 24 percent a year.

—A substantial share of inflation for the poor stems from the "other" category of consumption; that is price increases not directly related to food, rent, energy, or medical care.

To recapitulate: to understand the effects of higher energy prices on the poor it is necessary to understand how energy costs are related to housing costs in general and how increases in energy costs are related to overall inflation and, especially, the relative inflationary effects upon low-income groups.

Beyond these price and cost issues is the issue of housing demand. A typical model of housing demand would involve the observed or implicit rental rate of housing services; the prices of complementary and substitutable goods; services such as energy, utilities, transportation; goods and services for maintenance and upkeep; and income.

The implicit rental rate for housing services would incorporate such housing characteristics as location, climate, space, and neighborhood effects.

47. Ibid.; also, simulation of components of the consumer price index, 1974–80, for all consumers.

In principle, if the relative price of a complementary commodity such as energy should increase, then the demand for housing services could decrease in the following different ways: changing location and climate; changing space; and changing neighborhoods. The first two could subsequently influence the demand for energy. To the extent that low-income households shifted their housing demand to housing in which the cost of energy is lower, the increased demand for smaller, lower-standard housing or housing located in a warmer region would, given an adequate supply, tend to increase the rental rate of these dwellings. The eventual net effect of higher energy prices and higher rental rates for less energy-consuming dwellings would tend toward an equilibrium in which households tended to stay put.

Within a given dwelling there are still substitution possibilities in response to higher energy prices; they include higher maintenance and upkeep expenditures to improve weatherization and higher expenditures for supplemental heating devices or fuels, such as wood, or even clothing and apparel to increase comfort at lower room temperatures.

In relation to overall spending for consumption the effects of higher energy prices upon housing for the poor and the elderly will to some extent depend on what happens to the prices of food, medical care, and transportation in addition to the rental rates of dwellings. The real issue from the standpoint of consumer demand is the possible substitution behavior of a full range of changes in relative prices and real incomes rather than changes in energy prices alone.

In fact, there is an unsubstantiated assumption that real energy prices will increase during the next ten years. This should be questioned. Oil import prices and retail gasoline prices have fallen in real terms since January 1981. There are analysts who believe that, decontrol of natural gas notwithstanding, the United States has only begun to experience the effects of long-run demand and supply responses upon oil and gas prices and that average real energy prices could continue to fall throughout the 1980s.

This is an intriguing prospect, and, given the present financial environment, it suggests that perhaps the United States should be considering the effects of the cost of capital, instead of the cost of energy, on low-income housing during the 1980s.

RICHARD F. MUTH

Energy Prices and Urban Decentralization

PERIODICALLY, certain events are heralded as the salvation of the central cities. During the late 1950s and early 1960s the federal urban renewal program was expected to reverse the tide of suburbanization. It soon became obvious that it would not. Around 1970 it was widely anticipated that demographics—the coming to young adulthood of the babies of the 1950s—would increase the demand for multifamily housing and central city living. Yet anguished cries of "undercount" by big-city mayors were as numerous following the release of preliminary 1980 census counts as they had been in 1970. Despite a few dramatic instances of so-called regentrification, the bald fact is that most central cities continued their relative declines in population in the decade just concluded, and many suffered population losses. Indeed, for 243 standard metropolitan statistical areas (SMSAs)—all those defined in the 1970 census—combined, central city population fell about 3.7 percent between 1970 and 1979.[1] Recently the dramatic increase in the prices of crude petroleum and other sources of energy is seen as producing a large-scale return to the city. Is this likely to take place?

I think not. For each of the possible effects I shall examine, changes in energy prices, though favoring the central city, are quantitatively small. The most obvious effect of increased energy prices is on the cost of commuting to work. Between January 1973 and April 1981, the real price of gasoline approximately doubled. Yet gasoline and oil accounted for only a fifth of the money costs of driving an additional mile in 1973. Other money costs are associated primarily with the cost of automobiles,

1. U.S. Bureau of the Census, *Statistical Abstract of the United States, 1980* (Government Printing Office, 1980), table 21, p. 18.

and between 1973 and 1980, the real cost of automobiles declined somewhat. In 1973 money costs were only about two fifths of the total cost of commuting an additional mile, moreover, the larger share being the costs of the time spent. When all these factors are combined, my calculations suggest that the real cost of commuting an additional mile rose only about 3.6 percent between 1973 and 1980. To be sure, such an increase would tend to inhibit further decentralization. My analysis suggests, however, that it would do no more than offset half a decade's decentralization arising from the growth of real income alone.

Other energy price effects are also likely to be small. The prices of energy used for household fuel and similar purposes have risen much less than gasoline prices. Energy used in the home, moreover, is only about one eighth of the total expenditure on inputs used in producing housing. It appears to me that the real increases in energy prices that have been observed since 1973 would reduce the demand for housing no more than 2 percent. Such a decline would have little effect upon the land area needed to house a given population.

The energy price effect associated with type of structure is also likely to be small. Single-family dwellings use almost exactly twice the energy used in otherwise comparable dwellings in structures that contain five units or more. Even so, energy accounts for a relatively small share of the resource cost of housing produced by one-unit structures—less than 14 percent. Increases in the price of household fuel since 1973 would have raised the relative price of housing in single-family dwellings only 1 percentage point more than the relative price in multifamily dwellings. Similarly, multifamily dwellings make up about half the housing stock in central cities but less than a quarter in the suburbs. Yet my calculations suggest that the increase in household fuel prices since 1973 would increase the housing price advantage of central city dwellings over suburban dwellings attributable to energy only 0.3 percentage points, from 1.8 to 2.1 percent.

All my calculations, then, suggest that the increased real cost of energy is likely to have little effect on urban spatial structure. In considering the issue, it is well to remember that urban decentralization is not a phenomenon that is peculiar to the post–World War II United States. Rather, all the available evidence suggests that decentralization has been going on since the nineteenth century—and perhaps earlier—and in cities of Western Europe as well. To be sure, decentralization has apparently proceeded most rapidly in periods of sharply declining costs of transporting people and goods, and increases in this cost may be expected to slow it. Yet increases in urban size and rising real incomes have also contributed significantly to decentralization. It is by no means likely that the end of an era of growth has been reached, either in the size of urban areas or in the well-being of their inhabitants.

The Costs of Commuting

Probably the most important effect of higher energy prices is that on the cost of commuting to work. Commuting costs are in turn related to the distribution of population within urban areas.

Urban Population Distribution

The principal focal point of most urban areas is the central business district (CBD). Though less important than it once was as an employment location, the CBD typically contains considerably more employment per unit of land than any other part of the urban area. Workers employed in the CBD, moreover, do not live solely in an immediately surrounding residential ring. Rather, they commute to the CBD from residences located throughout the urban area. In 1970, about 8.5 percent of all workers who lived and worked in the same SMSA of 100,000 persons or more were employed in the CBD of its central city—or one of its central cities. The percentage was higher—12.4—among workers living inside a central city than among those living outside it—5.2.[2] Since SMSAs are defined as combinations of counties, some of which have substantial rural parts, the last figure is probably a substantial understatement of the percentage of suburban workers employed in the CBD.

CBD commuters from residential locations at greater distances bear greater costs in getting to and from work than do those who travel shorter distances. Calculations that will be presented later suggest that it cost roughly forty dollars a year in 1973 dollars to commute an additional mile by automobile. If a household were to spend the same amount for housing of a given quality, it would be less well off than a household that included a worker in the CBD who lived one mile closer to the downtown area.

2. Bureau of the Census, *1970 Census of Population*, vol. 1, *Characteristics of the Population*, pt. 1, *United States Summary* (GPO, 1973), table 242, p. 830.

It would thus be in the interest of more distant households to try to move closer to the CBD. In the process, the implicit rental value of owner-occupied dwellings closer to the CBD would tend to rise, and those of dwellings farther from it would tend to fall. It can be shown that for households of given tastes and income to be equally well off in different locations in relation to the CBD, the percentage difference in the price of housing services—that is, the rental value of identical dwellings—between any two locations must be equal to the difference in commuting costs between them divided by housing expenditures per household.[3] By the term "housing services" I mean simply whatever bundle of satisfactions a household receives by inhabiting a dwelling unit.

It is frequently asserted that the foregoing analysis overemphasizes the influence of the CBD on the distribution of population in urban areas. This is partly because only a minority of the workers in an area are employed there and partly because it neglects the influence of other employment centers. So long as significant numbers of workers are employed in the CBD and commute to it from locations throughout the urban area, however, those commuting greater distances must be compensated for their longer trips. Workers who do not commute to the CBD receive lower wages than those who do, and locally employed workers at progressively greater distances from the CBD receive progressively lower money wages to compensate them for the fact that housing prices are lower at greater distances from the downtown area of the city. More important, as I shall argue later, the model based on the influence of location in relation to the CBD predicts changes in urban population distribution from 1950 to 1970 quite accurately. I therefore infer that the suburbanization of jobs or alleged changes in tastes for suburban living in preference to central city living have been relatively unimportant influences on postwar urban decentralization in the United States.

If the rental value of identical dwellings declines with distance from the CBD, commuters living farther away would occupy somewhat bigger and better dwellings, though they would probably spend roughly the same dollar amount per year on housing. The principal effect, though, of commuter transport costs and housing prices that decline with distance from the CBD is in the way dwellings are built. Building materials and

3. See, for example, Richard F. Muth, *Cities and Housing: The Spatial Pattern of Urban Residential Land Use* (University of Chicago Press, 1969), pp. 22–23.

the labor to construct and to operate dwelling units probably cost pretty much the same throughout an urban area. Land rentals, however, vary in response to variations in housing and other demands for urban land, since land is not transportable. Since producers of housing, including the owners of houses, who are, in effect, their own landlords, receive lower monthly rentals for identical dwellings located farther from the CBD, they offer less for land located farther away. Indeed, since the cost of raw land is probably about 10 percent of the cost of providing housing space and since land rentals must absorb the whole of the decline in housing rentals, land rentals would decline about ten times as rapidly with distance from the CBD as the space rental values of dwellings.

For a variety of reasons, less land is used per dwelling where land is more expensive. An important source of economizing on land is in the type of structure built. Close to the center of many large U.S. cities, where land is relatively expensive, many dwellings are provided in elevator apartment buildings of ten stories or more. Farther away, garden apartments and row houses are more prevalent, while in the outer parts of urban areas most dwellings are in single-family, detached structures. Given types of structures typically occupy more land, moreover, the greater their distance from the downtown area. Thus, the value of housing services per unit of land tends to decline with distance from the CBD. The extent of the decline depends both upon what fraction of housing costs land costs are and how readily land can be substituted for nonland capital in producing housing services.[4] My best estimates of the two factors imply that the relative rate of decline in the value of housing services produced per unit of residential land is about eight times as great as that of dwelling rental values.[5]

Population density—persons per unit of land—is equal to the value of housing services produced per unit of residential land divided by the per capita expenditure on housing services. The considerations discussed above suggest that, apart from variations in average size of household, population densities decline at about the same rate with distance from the CBD as the value of housing services produced per unit of residential land. My calculations suggest that the rental values of identical dwellings declined at the rate of about 2 percent a mile during the 1970s, so population densities declined at roughly 16 percent a mile. The latter is

4. Ibid., pp. 54–55.

5. See Richard F. Muth, "Numerical Solution of Urban Residential Land-Use Models," *Journal of Urban Economics*, vol. 2 (October 1975), pp. 314, 318–19.

almost exactly the median of the estimates of percentage rate of decline in population density for fifty U.S. cities in 1970 made by James B. Kau and Cheng F. Lee.[6] Though considerable variation exists among these cities, for thirty of the fifty the estimated percentage rate was in the range of 0.1 to 0.2.

Consider now what happens as a city grows in population. With a larger population, a greater quantity of housing is demanded throughout the city. Consequently, housing prices rise everywhere, and producers of housing offer higher rentals for residential land. At the city's old edge, land that had been equally valuable whether used for urban or rural purposes now has a higher value if devoted to housing and other urban uses. In time, then, open fields will be converted to urban land. At the same time, land will become more expensive in the developed city area. When rebuilding takes place, less land will be used in relation to materials and labor in producing housing, and the city will gradually become more densely built up. On the basis of the foregoing analysis, the land area of the city would be expected to grow and population density to become greater as its population grew.

Obviously, larger cities have greater population densities and occupy larger land areas than smaller ones. This is true whether comparisons are made among different cities at a given time or in the same city in the course of time. Empirically, however, there are strong tendencies for the rate of decline in population density with distance from the center to become smaller as cities grow larger, but there is no really good theoretical explanation for these tendencies.[7] Indeed, to the extent that the volume of commuter traffic to and from the CBD grew, commuter transport costs would rise because of a slower rate of travel and cities would therefore tend to become more, not less, compact. Of course, as the city grew in size it might support more and larger subcenters for shopping and for retail and service employment. While this would reduce the relative importance of the CBD, it is hard to see how the CBD would become absolutely less important or that commuter transport costs would decline. Another explanation might be that housing output is more rapidly expandable in the parts of cities that are more distant from the center. First, a higher proportion of land might be vacant, awaiting

6. "A Random Coefficient Model to Estimate a Stochastic Density Gradient," *Regional Science and Urban Economics*, vol. 7 (March 1977), pp. 169–77.

7. Muth, *Cities and Housing*, chap. 7; and Edwin S. Mills, "Urban Density Functions," *Urban Studies*, vol. 7 (February 1970), pp. 5–20.

development. Second, though more square feet of land are used per dwelling in the outer parts of cities, the fraction of housing costs accounted for by land is smaller there. The output of housing is thus more readily expandable where land, which is in fixed total supply, is relatively inexpensive.

Rising average incomes have also caused cities to decentralize, and, unlike the effects of population growth, their effect is readily interpreted with reference to the foregoing analysis. Until the early 1970s, real earnings in the United States grew at a rate of about 2.5 percent a year or slightly more than 28 percent a decade. The increase in real earnings increases the cost of time spent in travel and hence the costs of commuter transport. The time cost is not the entire cost of commuting, however, so the latter would rise less than in proportion to earnings. With rising real earnings and therefore rising incomes, however, the value of housing services consumed by a typical family would rise as well. It is well known that consumption expenditure has risen in proportion to income in the long run. Apart from some fluctuations caused by the Great Depression and World War II and its aftermath, housing space rental expenditures have risen about in proportion to total consumption. These expenditures were 15.1 percent of total personal consumption in 1929, 14.0 percent in 1957, 15.2 percent in both 1965 and 1973, and 16.0 percent in 1979.[8] As a result of these changes, commuter transport costs would tend to fall in relation to housing expenditures as incomes grew, and urban areas would become more decentralized.

The rise in average income also increases the total quantity of housing demanded by a given population, which causes housing prices and the output of housing services to rise throughout the urban area. As does a rise in population, then, a rise in income causes the urban area to grow in size. Because commuter transport costs decline in relation to housing expenditure, moreover, housing output and population grow at more rapid rates in the outer parts of urban areas than in their older, more centrally located parts. Indeed, though housing prices and output close to the center rise, the quantity of housing services consumed by the average family rises as well. For this reason, total population toward the center may actually decline.

A decrease in population toward the center, however, is certain to

8. U.S. Department of Commerce, *Historical Statistics of the United States*, pt. 1 (GPO, 1975), pp. 316–19; and *Survey of Current Business*, various issues.

occur as a result of developments that reduce the costs of commuter travel. Particularly important among these are those that increase the speed of travel, since the time cost of commuter travel is more than half the cost of commuting. It appears for example, that in the mid 1950s, before the urban freeway had been built in most cities, the typical speed of rush hour commuting was about 20 miles an hour.[9] The urban freeway probably doubled this average rate of travel, though I don't know of any studies that document this point. Faster commuter travel, of course, reduces the time costs of commuting. Doubling the speed of travel would halve the time costs of commuting and, as can be seen by doubling the time costs for 1973 shown in table 4, would reduce the total cost of commuting an additional mile by about three eighths.

With the decline in commuting costs, housing prices, the output of housing services, rent of land, and population densities all would have to decline less rapidly with distance from the CBD. Housing prices, however, are fixed at the edge of the city by nonurban land values and construction costs. If the city had remained the same size but housing prices had increased less rapidly toward its center, housing prices would be lower throughout the city. Lower housing prices everywhere, in turn, would mean less intensive development of residential land and a smaller output of housing. To house the same population as before, therefore, the urban area would have to grow. To accomplish this growth, prices would rise at the city limits. But they could not rise so far that they remained at their former level in areas close to the center, for if they did, total housing output would be greater than it had been. Consequently, the reduction in commuting costs brought about by improvements in transport would lead to a decline in housing prices, housing output, and population in areas close to the center of an urban area and to increases in those parts of it close to the edge. Rising commuting costs caused by higher energy prices would have the opposite effect.

Empirical Evidence

It has long been known that, to a first approximation, population densities in urban areas decline at constant percentage rates with distance from the center. The percentage rate of decline is called the density

9. J. R. Meyer, J. F. Kain, and Martin Wohl, *The Urban Transportation Problem* (Harvard University Press, 1965), p. 81.

Table 1. *Average Central Population Densities and Density Gradients in Six U.S. Cities, Selected Years, 1910–63*[a]

Year	Central density (thousands per square mile)	Density gradient (proportionate decline per mile)
1910	72.5	0.96
1920	80.2	0.94
1930	59.1	0.73
1940	54.0	0.67
1948	45.2	0.57
1954	36.4	0.46
1958	32.1	0.41
1963	28.4	0.36

Source: Edwin Mills, "Urban Density Functions," *Urban Studies*, vol. 7 (February 1970), p. 13.
a. The six cities are Baltimore, Denver, Milwaukee, Philadelphia, Rochester, and Toledo.

gradient. In a variety of studies, both population densities at the center and the percentage rate of decline of urban population densities have been estimated. Probably the most comprehensive set of estimates is the set made by Mills.[10] Table 1 shows average values of population densities at the city center and density gradients for six U.S. cities for the period 1910–63. The latter declined persistently throughout the whole of the period, the former also following 1920. The decline in the percentage rate of decrease in density was most rapid, however, during the 1920s and in the post–World War II period. It was during these years, of course, that commuting by automobile increased most rapidly.

Other evidence suggests that urban decentralization is a more or less continuous phenomenon and not at all unique to the postwar United States. Estimates of the relation of population density to distance from the center for Chicago during the period 1860–1950 show a steady decline in the percentage rate of decrease in density throughout the ninety-year period.[11] The greatest decreases in the density gradient and the most marked declines in central density occurred during the 1880s—the period when rapid transit facilities were built—and during the 1920s—when automobile transportation became widespread. Colin Clark's estimates for European cities in the late nineteenth and twentieth centuries show

10. Mills, "Urban Density Functions."

11. Halliman H. Winsborough, "A Comparative Study of Urban Residential Densities" (Ph.D. dissertation, University of Chicago, 1960).

similar declines in the percentage rate of decrease in population density.[12] Estimates of population density functions for a large number of U.S. cities have been made by Muth for 1950 and by Kau and Lee for 1970.[13] Of the forty-four urban areas for which density gradients were estimated in both studies, the 1970 estimate was numerically smaller in thirty-nine.[14]

All the evidence cited above suggests that urban decentralization has been a widespread phenomenon in the Western world. That it has proceeded most rapidly during periods of improvement in transportation lends support to the hypothesis that decentralization is greater where commuting costs are lower. Examination of variation in my estimated density gradients for 1950 also supports this hypothesis.[15] The estimated density gradients varied inversely with car registrations per capita, which I used as an inverse proxy for automobile commuting costs. This remained true when the possibility of simultaneous determination of density gradients and car registrations was allowed for. Car registrations, moreover, were inversely related to the fraction of the population of an urbanized area residing in the central city and were directly related to the land area occupied by the urbanized area. These, of course, are precisely what the theory outlined earlier would predict. The density gradient estimates, moreover, were directly related to the population of the central city and inversely related to the land used by the urban area, also as the theory would predict. There is sound empirical justification, therefore, for examining the effect of high gasoline prices on the rate at which housing prices decline with distance from the centers of urban areas.

Gasoline Prices, Commuting Costs, and the Rate of Decline in Housing Prices

Everyone is familiar with the dramatic increase in nominal energy prices since mid 1973. In assessing the effect of increases in energy prices

12. Clark, "Urban Population Densities," *Journal of the Royal Statistical Society*, series A, vol. 114 (pt. 4, 1951), pp. 490–96.

13. Muth, *Cities and Housing*, chap. 7; Kau and Lee, "Random Coefficient Model," pp. 173–75.

14. I leave it to the skeptical reader to calculate the probability of such a divergence or a greater one on the assumption that increases and decreases are equally likely.

15. Muth, *Cities and Housing*, chap. 7.

Table 2. *Consumer Price Indexes, Selected Years, 1973–80*

Item	*1973*	*1976*	*1980*	*1980/1973*
1967 = 100				
All items	133.1	170.5	246.8	1.85
New automobiles	111.1	135.7	179.3	1.61
Gasoline	118.1	177.9	419.3[a]	3.55
Fuel and other utilities	126.9	182.7	278.6	2.20
Deflated by all items				
New automobiles	83.5	79.6	72.6	0.87
Gasoline	88.7	104.3	169.9	1.92
Fuel and other utilities	95.3	107.2	112.9	1.18

Source: U.S. Bureau of Labor Statistics, *Monthly Labor Review*, various issues.
a. For April 1981 (see text).

on the cost of commuting an extra mile, automobile transport is of overwhelming importance. In 1970, in only 24 of the 125 SMSAs with populations of 250,000 or more did as many as a third of the workers living in the central city and working in the CBD use public transport; in only 9 of these SMSAs did as many as a third of suburban CBD workers use it.[16] Table 2 shows some of the components of the consumer price index (CPI) that are pertinent to an evaluation of the effects of the increase in energy prices on automobile commuting costs. Because gasoline prices were fully deregulated only in early 1981, I have used the April 1981 price in place of the annual average for 1980. As table 3 suggests, nominal gasoline prices have risen about three and a half times since 1973, while the prices of fuel and other household energy utilities rose about two and a half times. When compared with the all-items index, the increases in energy prices are less striking, yet real gasoline prices have almost doubled since the end of the period preceding the series of oil shocks administered by the Organization of Petroleum Exporting Countries (OPEC). As table 3 shows, a doubling of real gasoline prices would increase the real cost per mile of operating an automobile about 16 percent. If the question is the effect of rising energy prices alone on the distribution of urban population, the last would be the correct figure to use.

It has been suggested to me that this figure both overstates and understates the increase in the money costs of motor fuel. It is an overstatement, it is asserted, because it neglects the rising fuel economy

16. Bureau of the Census, *1970 Census of Population*, vol. 1, pt. 1, table 364, pp. 1679–80.

Table 3. *Ten-Year Average Cost of Operating a 1976 Subcompact Automobile*
Cents per mile

Year and deflator	*Gasoline and oil*	*Various operating costs*[a]	*Total*
1. 1976	1.80	6.89	8.69
2. 1973	1.19	5.64	6.83
3. 1980	4.24	9.10	13.34
4. 1980 in 1973 dollars (1980 automobile prices)	2.29	4.91	7.19
5. Line 4 divided by line 2	1.92	0.87	1.05
6. 1980 in 1973 dollars (1973 automobile prices)	2.29	5.64	7.93
7. Line 6 divided by line 2	1.92	1.00	1.16

Sources: 1976, L. L. Liston and C. A. Aiken, *Cost of Owning and Operating an Automobile, 1976* (Federal Highway Administration, Office of Highway Planning, 1976), table 4, p. 15; 1973 and 1980, author's calculations.
a. Includes depreciation, repair and maintenance, replacement tires, accessories, sales tax, and taxes on gasoline and tires.

of cars since 1976. I think this assertion mistaken for two reasons: First, the calculations shown are based on the costs per mile of subcompact cars, and much of the increase in fuel economy has been the result of changes to automobiles of types that use less fuel per mile. Second, the fact that less fuel is consumed when its price rises is part of the response to a price increase and has nothing to do with the size of the price increase itself. The actual rise in price, it has alternatively been suggested, is not a correct measure of the effect of increases in the cost of energy on the costs of operating an automobile because of the possibility of interruptions in supply. Such an argument was valid during the period in which gasoline prices were subject to controls. Now that gasoline prices have been decontrolled, however, future interruptions in supply will raise gasoline prices, not cause waiting lines at gas stations. It may well be that automobile commuters consider the possibility of future increases in gasoline prices in their evaluation of automobile commuting costs, but I have no way of estimating the size of this effect.

The real money costs of automobile commuting have risen far less rapidly, however, than the increase in gasoline prices alone would suggest. This is the case because the costs associated with driving an automobile have declined since 1973. If what is likely to occur in the future is to be considered, rather than the hypothetical effects of energy price increases alone, the effect of the decline in the real costs of driving an automobile should be considered as well. The first line of table 3 shows elements of the money costs of commuting. From Federal Highway Administration (FHA) estimates of the ten-year costs per mile of operating a 1976 subcompact car, I have omitted items such as garaging, parking, and registration, which are independent of the number of miles a car is driven. Standard and compact cars, of course, have higher costs per mile, but if they are used for commuting such vehicles provide something more than pure transportation. The first line of table 3 indicates that petroleum costs were only about a fifth of the costs of operating an automobile in 1976. More than two thirds of the costs are those of depreciation, repair, and maintenance, which depend more closely on automobile prices themselves. Table 3, moreover, indicates that real automobile prices declined about 13 percent between 1973 and 1980. In the second and third lines of table 3, I have adjusted the costs of gas and oil and all other operating costs by the CPI component indexes for gasoline and for new automobiles, respectively. When the separate components are recombined, automobile operating costs almost exactly doubled in nominal terms between 1973 and 1980. When line 3 is adjusted for the decline in the value of money, though, the increase is much less dramatic. The cost per mile of 7.19 cents in 1980, when expressed in 1973 dollars, was only about 5 percent greater than its 1973 value.

When time costs are added, as in table 4, the increase in real commuting costs is smaller still. Taking total labor compensation—wages and salaries, plus other labor income—from the national income and product accounts for 1973 and dividing by total hours worked by full-time and part-time employees yields a total labor income of almost exactly $5.00 an hour, in 1973 prices.[17] The available studies suggest that commuters value their time spent in commuter travel at about 0.4 times their hourly wage rate, or $2.00.[18] Assuming a travel speed of forty miles an hour implies time costs of five cents a mile. Adding this to the money costs per mile of commuting yields estimates of total costs—money plus time—of 9.27 cents a mile in 1973 and 9.49 cents a mile in 1980, or an increase of 2.4 percent. Since there has been little, if any, increase in real hourly earnings since 1973, I have assumed constant real time costs

17. U.S. Bureau of Economic Analysis, *The National Income and Product Accounts of the United States, 1929–74: Statistical Tables* (GPO, 1977).

18. These are summarized in Gregory K. Ingram, John F. Kain, and J. Royce Ginn, *The Detroit Prototype of the NBER Simulation Model* (National Bureau of Economic Research, 1972), p. 69.

Table 4. *Cost of Commuting an Additional Mile, by Automobile, in 1973 and 1980 Prices*
Dollars

Item	*1973*	*1980 (in 1973 dollars, 1980 automobile prices)*	*1980 (in 1973 dollars, 1973 automobile prices)*	*1980/1973 (1980 automobile prices)*
Costs per mile				
Time[a]	0.0500	0.0500	0.0500	1.000
Money[b]	0.0427	0.0449	0.0496	1.052
Total	0.0927	0.0949	0.0996	1.024
Annual costs[c]	44.5000	45.5500	47.8100	...
Annual costs in relation to housing expenditures, space rent only	0.0247	0.0253	0.0265	...

Source: Author's calculations, based on table 3, as discussed in text.
a. Forty percent of average hourly earnings ($5; see text) at forty miles an hour.
b. Assuming 1.6 persons per car.
c. Ten one-way trips a week, forty-eight weeks a year.

in the calculations. Increases in energy prices during the 1970s may have contributed to the slowdown in the growth of real earnings. Such increases, however, lead to a once-and-for-all-time decline in real output, not to a reduction in the rate of its growth. Thus, the doubling of real gasoline prices in 1973 has been associated with an increase of less than 5 percent in commuting costs.

In table 4 the costs per mile are converted to annual costs of living and commuting a mile farther from the CBD by multiplying by 480. The last is the product of two trips a day, five days a week, forty-eight weeks a year. As will be seen in the next section, space rental expenditures in 1973 were almost $1,800 per household a year in 1973. The evidence also indicates that real income per household has remained roughly constant since 1973, so I have used the same real housing expenditure for the 1980 calculation. These figures, then, imply that housing prices in the typical U.S. city would decline about 2 percent a mile. They suggest further that the percentage rate of decline in housing prices would rise from about 2.47 to 2.53 percent, assuming a constant expenditure on housing services, because of the increases in automobile commuting costs since 1973, or about 2.4 percent. The increase in the cost of gasoline alone would lead to an increase from 2.47 percent to 2.65 percent.

Table 5. *Effect of a Decade's Growth in Income on Marginal Commuting Costs*[a]
1973 dollars

Item	*Initial*	*Final*	*Final/ initial*
Annual costs			
Time	24.00	29.31	1.221
Money[b]	20.50	20.50	1.000
Total	44.50	49.81	1.119
Annual costs in relation to housing expenditure, space rent only	0.0247	0.0227	0.9190

Source: Author's calculations, based on table 4, as discussed in text.
a. Assuming income growth at 2 percent a year.
b. 1973 prices.

To evaluate the practical significance of the last number, consider the effect of a decade of growth in real income on the percentage rate of decline in housing prices. The pertinent calculations are shown in table 5. Here it is assumed that real hourly earnings grow at a rather conservative rate—because they have increased very little if at all since 1973—of 2 percent a year, or about 22 percent throughout a decade. If so, the annual time costs of commuting an additional mile would rise by a like percentage, or from $24.00 to $29.31, in real terms. At given 1973 commuting costs, the total annual costs of commuting an additional mile would rise a little less than 12 percent. Assuming that incomes grew at the same rate and, as seems appropriate from my earlier analysis, that the space rental expenditures grew at the same rate, or to about $2,200, the percentage rate of decline in housing prices would fall from 2.47 percent to 2.27 percent a mile in the course of a decade, or somewhat more than 8 percent, because of the growth in real income alone. The effect of increased automobile commuting costs between 1973 and 1980, then, would be to slow decentralization somewhat, by roughly three years' growth in income—nine years if the decline in real automobile costs is neglected—but certainly not to reverse it.[19]

Indeed, I have heard it suggested that, because new dwellings will be built in such an energy-efficient manner, older areas will actually empty as people seek newly built suburban dwellings. This, too, it seems

19. Kenneth A. Small, in "Energy Security and Urban Development Patterns," *International Regional Science Review*, vol. 5 (1980), pp. 97–117, likewise concludes—though on the basis of rather different considerations—that increases in energy prices are likely to have little effect on the overall distribution of the urban population.

to me, is unlikely to occur. New construction takes place in disproportionate amounts in suburban areas, to be sure. But a glance at the decennial census of housing quickly shows that new construction takes place in the central city as well. So long as new construction occurs in both the central city and the suburbs, the relative prices of new dwellings in the two locations are influenced by differential commuting costs in the manner already discussed. If older dwellings in either location are less energy-efficient, their relative prices decline, so consumers are equally well off when living in older dwellings—spending more on energy but less for space—or in newer dwellings and the reverse.[20] Indeed, if the price of space in older dwellings were to decline sufficiently, these dwellings would be demolished and replaced with new, presumably more energy-efficient dwellings. In any event, persons inhabiting either old or new dwellings at different locations would have to be compensated for differences in commuting costs by differences in housing prices.

I have also heard it argued that some business firms will move to the suburbs to be closer to their workers because of the energy price increases of the past decade. Such an argument is almost impossible to evaluate because almost nothing is known about the determinants of the intrametropolitan location of business firms. Even if firms did so, however, it is by no means clear that the distribution of population between central city and suburbs would be affected. Central city firms are now net importers of labor from the suburbs, and some of those that moved might merely reduce aggregate net imports. So long as workers would still commute to the CBD in significant numbers, the distribution of population between central city and suburbs would be unaffected. I find it difficult to believe that the effect of energy price increases would be so strong as to reduce commuting to the CBD to an insignificant trickle.

Other Energy-Price Effects

There are two other avenues through which the effect of energy prices on urban decentralization might work. The first is through the consumption of housing space itself. If the rise in the prices of heating and other energy were to reduce the per capita consumption of housing significantly,

20. In the paper by Kevin Neels and Michael Murray in this volume similar conclusions are reached.

a given amount of housing production would house a larger number of persons. Urban areas would thereby be more compact than they would be otherwise. The other avenue is through differential effects of energy price rises. Energy use per comparable dwelling appears to be considerably greater in single-unit than in multiunit structures. The rise in energy costs, then, could conceivably increase the demand for housing in multiunit structures, which are more prevalent in the central city than in suburban areas.

Housing Consumption

To my knowledge there is no evidence on the response of the demand for housing consumption, defined as space rent, to changes in energy prices. One approach to the problem is to regard space, energy, and other current inputs as inputs into the production of a composite commodity broadly defined as housing. The response of housing space demanded would then depend upon a variety of factors. The greater the expenditures for energy in relation to expenditures on the bundle of commodities that consists of space, energy, and household operation, the greater this response. The response of the demand for space would also be greater the greater the response of demand for the composite bundle to changes in its price. Finally, the more easily the other items in the bundle could be substituted for energy without changing the consumer's satisfaction with the bundle the smaller the response would be.

Of all the elements just enumerated, the relative importance of energy expenditures is easiest to determine. Expenditures per household in 1973 on various aspects of housing suggest that the typical U.S. household spent just over $400 on energy for household use in 1973, about 12 percent of its total expenditure on housing broadly defined.[21] From a variety of comparisons made through the years, my best estimate is that the quantity of housing services demanded declines about in proportion to increases in their price. If, as many appear to believe, demand for housing is noticeably less responsive, my estimate of the effect of using energy prices is too large.[22] My own investigation suggests that there is little substitutability between capital and current expenditures in the

21. Calculated from data in Bureau of Economic Analysis, *National Income and Product Accounts*, 1929–74.

22. See, for example, A. Mitchell Polinsky, "The Demand for Housing: A Study in Specification and Grouping," *Econometrica*, vol. 45 (March 1977), pp. 447–61.

production of housing.[23] Using these figures, a 10 percent increase in energy prices would reduce the demand for housing space about 1.2 percent. The 18 percent increase in the costs of fuel and other energy utilities shown in table 3 thus suggests a decline of about 2 percent in the demand for housing space. Though the precise effect could be considerably different, the fact that energy prices account for roughly 12 percent of housing expenditure broadly defined implies that the recent increases in real energy prices will have little effect upon the demand for housing space and expenditures on it. Consequently, little effect upon urban structure can be expected from this source as well.

Type of Structure

The only other avenue of which I am aware through which energy prices might affect urban structure is type of structure. It is fairly well known that single-unit structures, which predominate in the suburbs, use more fuel than dwellings in multiunit structures. What is surprising to some, perhaps, as it was to me, is the extent of the difference. In a thorough recent study in which data on a very large number of characteristics of dwellings and households were used, the pure effects of type of structure, holding other characteristics of housing fixed, were analyzed. The study suggests that single-unit dwellings use almost exactly twice the energy per unit consumed in comparable units in dwellings in structures having five units or more, whereas two-unit dwellings use 1.73 times as much per unit, three-unit dwellings use 1.45 times as much, and four-unit dwellings 1.15 times as much.[24]

What does this imply about the differential effect of energy prices on the price of housing services in different types of structure? According to the 1970 census, roughly 64 percent of dwellings in SMSAs were in one-unit structures, 36 percent in two-or-more-unit structures.[25] Making the extreme assumption that energy use is twice as great in the former as in the latter and accepting the $414 figure of the Bureau of Economic Analysis as typical of the average SMSA, energy costs were 13.9 percent

23. Muth, "Capital and Current Expenditures in the Production of Housing," in C. Lowell Harriss, ed., *Government Spending and Land Values: Public Money and Private Gain* (University of Wisconsin Press, 1973), pp. 65–78.

24. Calculated from Kevin Neels, "Families, Houses, and the Demand for Energy," Rand Note N-1542-HUD (Santa Monica, Calif.: Rand Corp., 1981), table 1, p. 14.

25. Bureau of the Census, *1970 Census of Housing*, vol. 1, *General Housing Characteristics: United States Summary* (GPO, 1971), pp. 1–53.

Table 6. *Effects of Different Types of Structure on Expenditures for Housing, Central City versus Suburb, 1973 and 1980*
Dollars

Item	*1973*	*1980*	*1980/1973*
Fuel expenditure per household			
Central city	381	451	. . .
Suburb	447	529	1.180
Total housing expenditure per household			
Central city	3,494	1,564	1.020
Suburb	3,560	3,642	1.023
Central city/suburb	0.982	0.979	. . .

Source: Using the proportion of single-unit dwellings in the central city as 0.512 and in the suburb as 0.773, calculated from U.S. Bureau of the Census, *Census of Housing*, 1970, vol. 1, pt. 1 (GPO, 1972), table 10, I assumed that single-unit structures use twice as much energy as structures having two or more units.

of the costs of housing services in one-unit structures but only about 7.5 percent for dwellings in multifamily structures. The 18 percent increase in real energy costs shown in table 2 implies increases in the unit price of housing services of 2.5 percent for one-unit structures and about 1.4 percent for dwellings in structures having two or more units. It would take an almost unbelievably large response for such a differential price change to have an appreciable effect upon the choice of type of structure.

Differences in type of structure also imply a very small differential effect upon the choice of central city housing or suburban housing. As table 6 indicates, though 77 percent of suburban dwellings were in one-unit structures, 51 percent of the units in the central city were. Again assuming that dwelling units in two-or-more-unit structures use only half the energy that single-family dwellings do, the next two lines show energy expenditures per dwelling in the central city and the suburbs at 1973 and 1980 fuel prices that are implied by differences in type of structure. The next two lines show total housing expenditure, defined as expenditure on space and household operation and assuming that non-energy expenditures are constant. These entries indicate that the increase in household energy prices between 1973 and 1980 would increase the price of housing services in the typical central city dwelling 2 percent, in the typical suburban dwelling 2.3 percent. Looked at somewhat differently, the central city advantage in the price of housing services arising out of savings in the costs of energy would increase from 1.8 percent to 2.1 percent because of the rise in real household energy prices between 1973 and 1980.

The foregoing calculations of course refer only to averages for the country as a whole and probably do not reflect the effect of rising energy prices on the demand for housing by every group of consumers taken separately. In some regions—New England is an example—home heating costs are a larger share of total expenditures, they have risen more than in others, and the effects on the demand for housing are probably greater there. Or, to take another example, low-income households tend to spend larger shares of their incomes for heating and other household utilities than higher-income households. Not only are the demands for housing on the part of low-income consumers therefore affected more strongly, but the real incomes of low-income persons are reduced more by the rise in home heating prices. From the calculations just presented, however, even if the effects for some groups were several times as large as for the average consumer in the United States, the total effect would still be small. The only reasonable conclusion to be drawn from these comparisons, it seems to me, is that energy price increases so far could have only a negligible effect upon the choice of residence in the central city over the suburbs.

Comment by Richard D. Morgenstern

Richard F. Muth's paper is an interesting piece that addresses a pointed question: How will the energy price increases of the past eight years affect the tendency toward urban decentralization, a phenomenon that has been observed in this country for the past fifty years or more and that may, as Muth notes, underlie European development from as long ago as A.D. 100? His answer, which appeals to me intuitively, has also been advanced by others in this audience. Recent energy price increases are, by themselves, likely to serve as only a modest brake on the tendency to decentralize. Because the demand for energy is what economists call a derived demand—in the aggregate accounting for 7–8 percent of GNP—dramatic changes in its price will, in general, affect final demand by only a small amount.

Muth documents his analysis with an interesting set of calculations, and he finds, for example, that after accounting for the costs of both money and time, the real cost of commuting an extra mile has increased only 3.6 percent since 1973. This is barely enough, he says, to offset half a decade of decentralization arising from his perpetual engine of decentralization, the growth in real income. He does not, of course, address the link between energy price increases and the dismal growth in real income during the past decade—but to criticize his paper for that would probably be unfair.

Specifically concerning housing, Muth argues that since energy expenditures account for only about an eighth of the total costs, recent increases in energy prices are not likely to reduce the demand for housing more than 2 percent. He notes further that despite the fact that single-family units, more common in the suburbs than in the central cities, consume considerably more energy per unit than typical or even comparable multifamily units, recent energy price increases will no more than offset the general decline in the demand for housing in central city units.

I find Muth's analysis solid. I have only modest quibbles with his calculations, and I generally agree with his conclusion that recent increases in energy prices are likely to have only small effects on urban spatial structure. Apart from my own views on the process of decentralization, which I see as dominated less by real income effects than by changing tastes and the growing employment opportunities in suburban areas, my general criticism of the paper stems more from the questions that are not asked than from the ones that are. Before taking up these questions, however, I should like to comment briefly on a number of Muth's calculations.

Technological Change in Automobiles

Initially Muth divides total commuting costs into two parts—40 percent money costs and 60 percent time costs. The literature on the valuation of time is, to be sure, controversial, and the average-income approach that Muth uses is but one method—and not an unreasonable one. The problem, however, is really with the money costs: Muth may, in fact, actually be *overstating* the effects of higher fuel prices. He notes that gasoline and oil constitute 20 percent of the money costs, or 8 percent of the total costs of commuting. The doubling in real terms of the prices of gasoline and oil between 1973 and 1978 is offset in part by the decline in the real prices of automobiles—depreciation, maintenance, and repair—so the net increase in money costs is only 8.5 percent. When weighted by the time costs, unchanged in real terms, the net effect of this increase in money costs is an increase in total commuting costs of only 3.6 percent.

These calculations are based on the costs per mile of operating a 1976 subcompact car. In fact, the greatest single energy-related change in the U.S. economy is probably the increase in miles per gallon of the automobile fleet. Between 1973 and 1980 the average miles per gallon obtained by new cars increased about 40 percent—and it is expected to increase much more in coming years. Since half the fleet is replaced every six years and new cars are driven more than older ones, the average fleet efficiency has probably risen at least 25 percent since 1973. A 25 percent increase in fuel efficiency of the entire fleet—not only of new cars—yields corresponding reductions in real fuel costs per mile. In combination with the decline in charges for depreciation, maintenance, and repair, this produces *no change* in the real money costs of commuting. Since time costs were posited not to increase in real terms, the net effect is that real commuting costs have probably not risen at all since 1973, rather than the 3.6 percent projected by Muth.

The Distribution of Heating Costs among Households

A second issue I would raise is related to the effect of fuel costs on total housing costs. Muth has used a Bureau of Economic Analysis index of electricity, gas, coal and fuel oil. While I agree that this is probably the single best measure, the regulated price of two of these fuels, electricity and natural gas, may mask significant considerations concerning distribution. The price of fuel oil—which in 1979 was the main heating fuel of about 20 percent of all the households in the country and of about half the households in the Northeast—has almost doubled since 1973. In contrast, delivered prices of regulated fuels, such as natural gas and electricity, have increased more slowly. In the future, increases in delivered prices of natural gas outside the producing states are likely to outpace the others. The real point is that the average increase in fuel prices masks some important variations. Thus, while correct on the average, Muth's estimates may systematically underestimate the effects that energy prices increases have had on the demand for housing by a large number of households and, since oil is used more widely than gas in outlying areas, on the differential effects between urban and suburban areas. In time, of course, decontrol of gas prices will eliminate much of the variation in delivered fuel prices, but not soon enough to prevent a significant distributive effect among households. This has been documented in a number of recent studies. The willingness of homeowners to spend up to $3,000 or more to convert from oil to gas is but one indication of the quantitative importance of this phenomenon.

The Availability of Fuel

While the memory of gasoline lines was still alive in many people's minds and the prospect of future interruptions in supply seemed very much of a reality, we would all have agreed that living in the distant suburbs carried a somewhat greater risk that commuting costs—in both money and time—would rise dramatically in the event of a serious interruption. Various attempts have been made to quantify this effect, which, in another context, has been labeled the supply-interrruption premium. This premium includes an assessment of the probability of a serious interruption. In November 1980 the Department of Energy estimated that because of the insecurity of supply the value of a marginal barrel of imported oil was four to ten dollars higher than its price, or about ten to twenty-five cents a gallon. Even if these estimates are accepted, however—and they would have been much lower a year later—the availability factor, which had been heralded by some, would have only a marginal effect on Muth's calculations and thus would not lead to significant spatial reallocation.

Returning to the basic issue, I repeat that I find myself in general agreement with Muth's conclusion that the rises in energy prices in recent years, including nonprice effects, if they exist, will probably not greatly affect the spatial allocation of economic activity within metropolitan areas. But since we are trained to be critical rather than laudatory—and since our chairman has invited us to broaden our horizons and thereby set a context for discussion—let me take a shot at it.

If the spatial allocation of resources within urban areas is not the problem, then what is? At last year's conference we spent some time describing the extent of energy price rises and we focused on the distributional effects of these price rises. Essentially, we looked at the question of which group of individuals, industries, or regions endured the greatest losses, at least implicitly, and which enjoyed the greatest gains as a result of the energy price hikes.

Today we want to address the question of how these different groups have responded to the price changes. Every economist knows that in response to rising prices less of an item is demanded. In fact, that is what is happening across the country—in spades. Even apart from the current recession, aggregate demand for energy per dollar of GNP or by any other measure of economic activity has fallen considerably in recent years. The principal mechanism by which demand for energy is reduced in the residential sector—and everywhere else—apart from turning the thermostat down, is substitution of other inputs, principally capital, for energy. This is commonly known as conservation of energy.

The reason for interest in this subject, it seems to me, is that while conservation of energy in the aggregate seems to be progressing rapidly, the pace is very different among different groups of the population. And since, by definition, we are talking about cost-effective conservation, the slow pace among some groups implies that resources are being used inefficiently. And if these inefficiencies are concentrated among groups that are already disadvantaged, the problem is of even greater concern.

Various data suggest that while opportunities for cost-effective conservation investments abound, many people are slow to respond. A recent survey in Maryland, for example, showed that only about 20 percent of the respondents planned to undertake important conservation investments during the next five years—despite the fact that energy audit data confirmed the existence of attractive opportunities for conservation investment in more than two thirds of the homes. Raymond J. Struyk has identified some of the age, income, and tenure characteristics of those who are least likely to make improvements in coming years. The question, it seems to me, is, How can the pace be accelerated so as to reduce the inefficient use of resources and, simultaneously, lend a hand to the groups penalized by the slow pace?

In the past, much attention was devoted to the barriers to investment in conservation in the residential sector—ranging from uncertainties in calculating the rate of return, through difficulties in obtaining financing, to a host of nonprice factors, such as discomfort with contractors. Rather than review these issues again, I should like to mention one of the more promising institutional developments designed to overcome some of these well-known barriers. Across the country utilities—principally electric utilities—are now becoming more active in promoting residential energy conservation. Especially in the high-energy-growth areas, utilities are demonstrating a growing interest in inexpensive ways of reducing demand, particularly peak demand. Some of the utilities in the Northwest and in California are offering a type of one-stop shopping service that goes beyond the audit program of the Residential Conservation Service. In some states the utilities are actually selecting contractors, financing the investments—sometimes at zero interest rate—and conducting a post-installation audit to see that the job is done right. Interest in these programs is remarkably high, and several studies are now under way to see how such schemes might work in different states.

In sum, I think that Muth has done a fine job of persuading us that the energy crisis, even assuming that it still exists, will not bring a great renaissance to the cities. What it may do, however, is create distortions and inequities in residential housing markets, as many people are slow to respond to the radically changed price regime. This conference, it seems to me, should be aimed at describing and analyzing these distortions and inequities, and at least raising the question of what remedies—including institutional changes that affect both single-family and multi-family units—might be appropriate.

KEVIN NEELS AND MICHAEL P. MURRAY

Energy and the Existing Stock of Housing

THE drastic increases in energy prices during the 1970s reduced the economic well-being of nearly all Americans significantly, but some sectors of the economy were hit harder than others. The American automobile industry, for example, suffered tremendous losses as consumers demanded smaller, more gas-efficient automobiles faster than the industry could alter its designs or its capital equipment. Suffering right along with the industry giants, although on a smaller scale, were thousands of individuals who found the resale value of their "gas guzzlers" plummeting.

The 1980s are likely to bring still further increases in real energy prices, and most Americans will again feel the bite. Some big losers of the 1970s, however, will fare better in the 1980s. The automobile industry, like many other producers, has altered its plans and its equipment to accommodate the higher energy prices, and drivers have sent many of the gas-guzzling cars to the scrap heap. Another breed of energy guzzlers is still with us, however—the immense stock of housing built before the surge in the real price of energy.

While rising energy prices will affect both newly constructed and existing housing, the burdens created will not be shared equally. Builders of new houses will take the rising cost of energy into account and alter their designs to reduce energy requirements. Owners of existing houses, however, will have fewer choices and will be more constrained in their responses. For them it will be harder to escape the burdens imposed by rising energy prices.

Nor can the owners and consumers of housing be expected to share the burden of rising energy prices equally. The impersonal mechanisms of the marketplace will decide who eventually bears the burden of higher energy prices. The competitive process, tempered by asymmetries of information between buyers and sellers, landlords and tenants, will act to mitigate the effects of higher energy prices when it is possible and to divide new costs of operation as they are incurred.

The effects of rising energy prices on the existing stock of housing depend upon patterns of energy use, technical possibilities for substituting other inputs for energy in newly constructed and existing housing, the importance of energy in the total cost of housing, and the responses of owners and occupants to the resultant pressures. In this paper we shall consider all these factors in an effort to determine what effects rising energy prices will have on existing housing in the years ahead.

Possibilities for Reducing Energy Use: Engineering Evidence

A primary factor in assessing the effects of rising energy prices on the housing market is the flexibility of housing technology and the ease or difficulty of substituting other inputs for energy. If the technology is highly flexible, an owner will respond to an increase in the price of energy by using something else in its place. The more flexible the technology of production, the more of the burden of rising energy costs the owner will be able to avoid. In an extreme case, he might be able to avoid the burden entirely. Rising energy prices would then have no effect on his costs. In contrast, if the technology is inflexible, a rise in energy prices would be translated directly into an increase in operating costs. In this other extreme case energy use would be fixed and the owner would simply have to pay whatever was necessary to obtain the required amount.

In a large number of engineering studies the technical possibilities for reducing the amount of energy required to operate a residential property have been examined. They provide direct evidence of the extent of the substitution possibilities in this area, background for the discussion ahead. Engineering studies of residential energy use are numerous and diverse, both in method and in emphasis. In some the statistical analysis of data on energy use drawn from a large number of actual dwellings is relied upon. In others careful measurements of energy use in particular buildings are made. In still others computer models are used to simulate the operation of limited numbers of idealized residential units. What they have in common is an interest in measuring the effect of particular actions on energy use.

Given the large number of engineering energy studies completed in recent years, it is difficult to select a small number for review. Presented here are the results of a sample. An examination of these findings will do much to convey an understanding of the current state of knowledge concerning a number of aspects of the subject.

Because the effects of a wide variety of possible actions have been measured in many of the engineering studies of residential energy use, the following discussion has been organized by action rather than by study, thereby making it easier to compare the results. We have tried to cast all the results of a particular action into a common format. This has required us to manipulate some of the figures presented in the source documents arbitrarily. We have attempted to preserve the spirit of the raw results, but we accept full responsibility for whatever errors may have been made in the translation.

Building Size

The amount of heat lost by an apartment through the shell of the building during the winter depends upon the amount of surface area exposed to the weather. In detached single-family dwellings, every exterior surface is exposed to the weather and susceptible to this type of energy loss. The presence of shared walls in multiunit buildings, in contrast, makes them inherently more efficient in their use of energy. In a number of studies the attempt has been made to measure the magnitude of the resultant savings, and surprisingly similar results have been obtained. They indicate that doubling the number of units in the building reduces energy use per unit approximately 13 percent. In the studies by Hittman Associates and AIA Research Corporation energy use was normalized to a per-square-foot-of-floor-space basis, and the tendency for apartment size to decrease as building size increased was thus controlled for.[1] Keyes made no such adjustment for building size in his

1. Hittman Associates, *Residential Energy Consumption: Multifamily Housing, Final Report,* U.S. Department of Housing and Urban Development, Report HUD-HAI-4 (Government Printing Office, 1974); and AIA Research Corporation, "Phase One/Base Data for the Development of Energy Performance Standards for New Buildings" (Department of Housing and Urban Development, Office of Policy Development and Research, 1978).

analysis, but he nonetheless came up with a lower estimate of the effect of increasing building size on energy use per unit.[2]

Insulation

The use of insulation as a way of reducing energy use in the residential sector has received a great deal of attention in recent years. A number of attempts have been made to measure the benefits that can be attained by installing insulation. The findings have been quite diverse.

Hittman Associates has estimated that the installation of insulation in the walls and ceilings of a residential structure would reduce energy use per unit 27 percent. In contrast, Keyes indicates that such an action would reduce energy use only about 11 percent. At 14 percent, the results of Hutchins and Hirst fall between these two extremes.[3] A similar diversity of opinion exists in connection with the effects of installing insulation below the ground floor. Hittman Associates found that this would reduce energy use 13 percent per unit. Hutchins and Hirst found that this would bring savings of no more than 3 percent, however.

The lack of agreement appears to arise from the fact that the various alternatives examined were poorly defined. The base cases in the studies by Hittman Associates and Hutchins and Hirst were units with "normal" amounts of insulation. The actions analyzed involved upgrading the quantities of insulation to supernormal amounts. Keyes, in contrast, analyzed the effects of going from no insulation to "some" insulation without defining clearly what this meant. There were also differences in the assumptions made about the types of structure being insulated and the amounts of other energy-saving equipment that were present.

It does seem clear from these studies that insulation in the walls and roofs is worth more than insulation below the first floor. The smallest estimate of the probable savings from installing insulation in the walls and roofs—11 percent—was still large in relation to the cost of installing insulation in a new structure.[4] In spite of the uncertainties surrounding the precise measurement of its effect, insulation still appears to be a

2. Dale L. Keyes and George R. Peterson, "Metropolitan Development and Energy Consumption," Working Paper 5049-15 (Urban Institute, 1976).

3. Paul F. Hutchins, Jr., and Eric Hirst, "Analysis of Single-Family Dwelling Thermal Performance," *Resources and Energy*, vol. 2 (September 1979), pp. 75–96.

4. Ibid.; Hutchins and Hirst analyzed the cost-effectiveness of installing insulation and found it to be the most economical of the options considered.

highly attractive way of economizing on the use of energy in the residential sector.

Window Alterations

Windows are an important element in determining the energy requirements of a building. In several studies the effects on total energy use of changing various characteristics of the set of windows on a building have been examined. It was found that substantial savings in energy were to be realized from the installation of storm windows. Neels estimated that the installation of storm windows would reduce energy use per unit 5 percent.[5] Keyes, in contrast, estimated that energy use would decline 37 percent. The estimates by Hutchins and Hirst fell between these two extremes. Neels's results on this point were weak statistically, and the higher estimates may be closer to the truth.

The double-glazing of windows appears to be less effective than the installation of storm windows. Hutchins and Hirst found that installing storm windows would reduce energy use 12 percent, while double-glazing would bring a reduction of only 6 percent. This finding was consistent with what would have been expected. Both storm windows and double-glazing reduce the amount of heat lost directly through the glass by conduction. Storm windows, however, have the added effect of reducing the infiltration of air around window sashes. These additional savings appear to be significant.

In the studies by Hittman Associates and Keyes, cited earlier, reductions in window area were found to reduce total energy use; estimates in the two studies were similar. Little should be made of this consistency, however, since a great deal of "massaging" was necessary before the two sets of results could be made comparable. The stronger results are probably those of Hittman Associates. They indicate that a reduction of 25 percent in window area would reduce energy consumption 8 percent.

Thermostat Setbacks

One of the earliest policy responses to the shortages caused by the 1973 Arab oil embargo was a call for a lowering of thermostat settings.

5. Kevin Neels, "Families, Houses, and the Demand for Energy," Rand Note N-1542-HUD (Santa Monica, Calif.: Rand Corp., February 1981), p. 14.

A number of attempts have been made to measure the savings that can be realized by turning thermostats down at night. Zabinski and Amalfitano found that lowering thermostat settings eight degrees between 10 P.M. and 6 A.M. would reduce the amount of fuel used for heating 17 percent.[6] In another experiment, Quentzel estimated the savings from an identical setback to be 9 percent.[7] The difference may be attributable to idiosyncrasies in the particular houses that were studied. In either case, the reductions were noteworthy.

Furnace Modifications

Some attention has been paid to the effects on energy use of changes in the characteristics of the heating equipment itself. Among those whose studies have been examined so far, Zabinski and Amalfitano paid the most attention to this possibility. They looked at the effects of varying the characteristics of both the burner and the nozzle. Replacement of the old burner with a modern flame-retention burner in the house they were studying reduced fuel consumption for space heating 15 percent. Replacement of the original nozzle with a slightly smaller one reduced fuel use 12 percent. The authors were careful in framing their conclusions, pointing out that these particular changes will not always have the same effect. Their results do suggest, however, that important benefits can be realized from a careful examination of existing heating plants and modification of them to suit them better to their job.

In a 1975 study the potential savings associated with the use of modular gas-fired boiler systems were investigated.[8] Most multiunit apartment buildings with central heat rely on a single boiler to heat water both for space heating and for direct domestic use. The size of this boiler must be determined by the maximum load that it will have to bear, which implies that most of the time it will be operating well below its rated capacity. Since most boilers attain their peak efficiency only when operating near full capacity, much fuel is wasted. A modular boiler system consists of several small boilers tied together in parallel. Extra boilers

6. M. P. Zabinski and A. Amalfitano, "Fuel Conservation in Residential Heating," *ASHRAE Journal*, vol. 18 (January 1976), pp. 41–46.

7. David Quentzel, "Night-Time Thermostat Set Back: Fuel Savings in Residential Heating," *ASHRAE Journal*, vol. 18 (March 1976), pp. 39–43.

8. G. E. Kelly and D. A. Didion, *Energy Conservation Potential of Modular Gas-fired Boiler Systems*, National Bureau of Standards Building Science Series 79 (GPO, 1975).

are brought into operation as the load on the heating system is increased. At any given time, in such a system, no more than one boiler will be operating below capacity. The result is much greater operating efficiency. Kelly and Didion estimated that at low operating loads the modular arrangement required 14 percent less energy than a more conventional single-boiler system.[9]

Applicability to Existing Housing

It is clear from this brief review that there are many possibilities even within the existing technology for reducing the amount of energy used by the residential sector. It is equally clear, however, that many of these techniques are applicable only to newly constructed housing. Owners of existing housing cannot change the size of their buildings without making uneconomically drastic modifications. They can modify windows and furnaces, but less freely than can builders of new housing. Occasionally, through conversions and mergers, they do change the number of units in a building, but this generally happens only when the initial sizes of the units are far from what consumers are demanding at the time. Conversion is not a possibility of which most owners can take advantage. This is especially true in the present instance, since the rise in the price of housing services will not induce more than moderate reductions in individual demands for housing services. Only in the cases of thermostat setbacks and insulation, then, do the two classes of owners have similar freedom of action. Because of the necessity of opening walls to install insulation, owners of existing housing obtain their freedom at a higher cost.

The Effects of Rising Energy Prices on Housing Consumers

Rising energy prices mean rising energy costs. This simple observation says nothing, however, about who will pay these costs. Much concern has been expressed about the burdens that rising energy costs will place upon one group, namely, housing consumers. We find that even under some extreme assumptions about the probable magnitude of the increase in real energy prices, the burdens placed upon housing consumers will

9. Ibid., table 3, p. 50.

be relatively modest. This happy outcome is the result of the way in which housing prices are determined and the overall flexibility of the production technology in connection with energy use.

A useful distinction is that between the physical stock of housing and the services derived from that stock during a given period; we shall call the former "the stock of housing" or "housing capital" and the latter "housing services." Housing services are produced by combining housing capital with other inputs—energy, land, and labor. The stock of housing is what owners possess; housing services are what consumers of housing purchase in each period. In the rental housing market, owners and consumers are distinct agents, but in the owner-occupant sector, the owners of housing capital are the consumers of the housing services produced with that capital. Nevertheless, it is useful to maintain the conceptual distinction between owners and consumers.

Producers of new housing will undertake new construction only if they can cover all their costs. As a consequence, if rising energy costs increase the cost at which housing services can be produced, consumers of housing services produced with new housing capital will have to pay a price—the price of housing services—sufficient to cover the new costs incurred by the producer.

Since producers of new housing can adapt to higher energy prices more easily than can owners of existing dwellings, producers of housing services who own existing units will incur higher costs after an energy price increase than producers of new units. It may be helpful to trace the consequences of these new higher costs.

Suppose a prospective buyer of a home comes into the market after the price of energy has risen. The buyer will reason, "If I buy the new house, I will be able to provide myself housing services at a monthly cost of X dollars, including the user cost of capital, the rental value of the land, the monthly value of labor expended, and the monthly cost of energy. I therefore won't buy the older house—which happens to offer the same flow of housing services—unless it costs me no more than X dollars a month." But if the older house is to match the price of housing services available in the new dwelling, the owner of the older dwelling must absorb the difference between the energy costs in the two dwellings; that is, he must suffer a capital loss.

Since the higher energy costs are incurred throughout the housing stock, the pressure on the value of existing capital will be felt throughout the market, not just at the margin where there is new production. The decline in the price of housing services in the existing units that are of higher quality—and are therefore competitive with new construction—will put similar pressure on owner-occupied units of somewhat lower quality, and so on, until consumers at the margin between renting and buying transmit the pressure from the owner-occupied market to the rental market. The process concludes by raising the price of housing services, and lowering capital values, throughout the market.

There is ample evidence that housing markets are tied together in the strong way described here. Johnson has found that sales prices of homes reflect the value of their energy-saving components.[10] While it is not entirely surprising that potential home buyers seek out relevant energy information, some readers will be surprised to learn that renters do something similar. Follain and Malpezzi estimate that contract rents in dwellings where tenants pay utilities are discounted considerably for higher-than-average utility costs.[11]

Perhaps the strongest evidence of the tight linkage among housing markets is the observed decline in the real price of rental housing services during the 1970s despite large increases in operating costs of such units. During the 1970s, the combination of tax laws, inflation, and slowly rising interest rates made it much cheaper than it had been for households to consume owner-occupied housing services. Operating costs rose but the rise was more than offset by capital gains and tax breaks. As a consequence, an unusually large number of renters became owners in the course of the decade—despite little or no increase in real incomes. In response to this effective price advantage enjoyed by owner-occupants, in order to compete, rental dwellings, despite higher real operating costs, had to be let at lower real prices. The price of housing services in the rental market declined sufficiently to dissuade the marginal renter from becoming an owner; these lower rents coupled with higher costs meant that owners of existing rental dwellings suffered significant capital losses.

Since the rising energy costs will strike all along the continuum of housing types, from rental units of the lowest quality to owner-occupied units of the highest quality, we can only believe that the market forces will act just as strongly in response to higher prices as they did in the face of shifts in the relative advantage of owning. As a consequence, the

10. Ruth Johnson, "Housing Market Capitalization of Energy-Saving Durable Good Investments," ORNL/CON-74 (Oak Ridge, Tenn.: Oak Ridge National Laboratories, 1981).

11. James R. Follain, Jr., and Stephen Malpezzi, "Dissecting Housing Value and Rent: Estimates of Hedonic Indexes for Thirty-nine Large SMSAs," Working Paper 249-17 (Urban Institute, 1980).

burdens that will be borne by housing consumers will depend upon the way rising energy prices affect production costs for builders of new housing. To understand this relation it is necessary to consider the way housing is produced.

The services derived from housing are produced in two distinct but related steps. First, a developer uses land, labor, and materials to build a dwelling. Second, a landlord or homeowner uses that physical structure plus other resources to provide shelter, comfort, and the like from day to day.

Energy has a largely indirect function in the production of the housing stock. Building materials are themselves produced with the aid of energy, so the price of energy influences construction costs through its effects on the price of building materials. On the other hand, energy is of central importance in the production of housing services. The comfort derived from residing in a particular dwelling is heavily influenced by the protection it affords against extreme temperatures.

How much energy is required to provide a particular level of housing services depends on the specific physical characteristics of the dwelling. A greater initial investment in the physical dwelling can enable the owner to conserve on the consumption of energy in the provision of housing services. This link between the housing stock and energy in the provision of housing services can be described by a production function for housing services:

$$H = f(K, E, N, L). \tag{1}$$

In this relation H is the flow of housing services, K is the physical stock of housing, E is energy use, and N and L are the amounts of labor and land, respectively, used in the production of housing services.

In connection with new construction, providers of housing services can choose a level of investment in housing capital, K, that minimizes the cost of providing a given level of housing services, given the prices of all factors. If energy and new capital were perfect substitutes in the production of housing services, when energy prices rose, producers of new housing could offer housing services at no increase in price. At the opposite extreme, if no substitution for energy were possible, the cost of providing housing services would rise along with energy prices. The degree of flexibility new producers have is crucial to residents, since the greater the flexibility, the less the price of housing services will rise as energy prices rise.

The evidence reviewed in the preceding section demonstrates that much can be done to reduce energy use in new housing. Econometric analyses of the production function for housing services also suggest that it is possible to design and build houses that are sparing in their use of energy. Neels has estimated the parameters of a translog production function, which is used in this paper, relating the quantity of housing services produced to input levels for capital, energy, labor, and land.[12] He estimates that there are extensive possibilities for substitution between capital and energy and also between labor and energy.

The advantage of using production function methodology to study housing technology is that it makes possible quantitative statements of the flexibility of that technology. In particular, from a production function it is possible to derive a cost function for housing services that takes as its arguments the prices of the four factors of production.[13] On the basis of the substitution possibilities provided by the technology, the cost function gives the lowest price that a profit-maximizing producer can charge for the housing services he produces and still cover his costs fully. Symbolically it is thus possible to write:

$$P_H = g(P_K, P_E, P_N, P_L), \tag{2}$$

where P_H is the price charged per unit of housing services and P_K, P_E, P_N, and P_L are the prices per unit of capital, energy, labor, and land, respectively. Coefficients of both the production function and the corresponding cost function are shown in the appendix to this paper.

Equations 1 and 2 afford us the opportunity to emphasize the difference between energy *use* and energy *efficiency*, which is often overlooked in policy discussions. If an individual wants to consume more housing services, more energy must be used. But if that additional housing service is produced at minimum cost, no energy inefficiency accompanies the increased energy use. If an individual can reduce energy use without increasing costs, however, not to do so would be energy-inefficient. Another way to phrase this distinction is to contrast "economizing" on energy use, or using energy efficiently to obtain a given level of output, and "conserving" on energy use, or cutting back on the goods and amenities produced with energy.

Assuming that the prices of other inputs remain constant in real terms,

12. Kevin Neels, "The Derived Demand for Energy in the Production of Housing Services (Ph.D. dissertation, Cornell University, 1981).

13. The relation between production and cost or price functions was first explored intensively by Ronald W. Shephard in *Cost and Production Functions* (Princeton University Press, 1953).

Table 1. *Effects of Energy Price Inflation on the Prices of Housing Services*

Real energy price index	Real housing services price index[a]
100.0	100.0
150.0	110.7
200.0	119.6
250.0	127.4
300.0	134.4
350.0	140.8

Source: Authors' calculations.
a. The prices of capital, labor, and land are assumed to remain constant in real terms. The prices of housing services are computed with the use of the price function described in the appendix to this paper.

we have computed the rise in the prices of real housing services associated with various increases in the real price of energy. The results are presented in table 1.

Rising real energy prices are likely to lead to both economizing and conserving. Providers of housing services will economize on the use of energy in production, and consumers, seeing new higher prices for housing services attributable to higher energy prices, will reduce their demand for housing, thereby conserving. Because the increases we predict in the prices of housing services are moderate, however, while both the price and income elasticities of housing demand are small, we believe it is the economizing that will dominate in determining the effect of energy prices on housing. As a consequence, in our simulations we restrict our attention to a fixed level of housing consumption.

We find that it is quite easy for builders of new housing to substitute other factors for energy. Our estimates show that if real energy prices were to double, real prices of housing services would have to rise less than 20 percent to allow producers of new housing to cover their new and higher costs. A tripling of real energy prices would force real prices of housing services up only slightly more than a third.

To place these changes in perspective, we should point out that during the period from 1972, before the explosion in energy prices, to 1980, the real price of fuel oil, which has led all other fuel prices, increased slightly more than 150 percent. A tripling of real energy prices is, therefore, far greater than anything yet seen. While a rise of 34 percent in real housing prices is not trivial, it is a small response, given the size of the input-price shock.

We should also point out a few things not implied by these results.

They do not imply that rents and occupancy costs are not going to rise. As the price of energy increases it tends to pull all other prices in the economy up with it. This general price inflation will tend to drive up rents for tenants and occupancy costs for homeowners. We are saying only that the differential effect is small; the price of housing services will not rise much in relation to other prices. Our results also do not imply that energy use per dwelling will necessarily decline. As incomes rise, households want to consume more housing. The fact that newly constructed dwellings are larger and more elaborately equipped than existing dwellings does not necessarily mean that they are less efficient—even if they use more energy. They may use more energy, but the increment in energy use could be small in relation to the higher level of housing services provided by the dwellings.

The Effects of Rising Energy Prices on Owners of Housing

The portion of the increase in the cost of energy for existing housing that is not borne by consumers will have to be paid by owners, in the form of reductions in the value of their property.

Estimates of the magnitude of the burden that rising energy prices will place on the owners of the existing stock of housing are subject to a great deal of uncertainty. Even apart from the difficulty of forecasting energy prices, there are two subjects on which we lack vital information. The first is the malleability of the existing housing stock; the second is the propensity of landlords and homeowners to take advantage of the opportunities presented to them by the technology of energy conservation.

We pointed out earlier that it is more difficult to reduce energy use in existing dwellings than in newly constructed dwellings. Hard evidence is lacking, however, on precisely how much more, and speculation has ranged widely. A few have likened the existing stock to the automotive gas guzzlers referred to at the beginning of this paper. Many believe that the efficiency of the existing stock can be improved but fear that doing so will require massive investments. Still others point to the huge savings that can be obtained from simple actions such as weatherizing and argue that large gains in efficiency are within easy reach.

A second source of uncertainty grows out of not knowing whether property owners will take full advantage of the possibilities for saving energy. Although economic theory asserts that landlords and homeowners will take any action that promises them a healthy return, many people have expressed concern whether, in fact, "enough" is really being done.

To get around these uncertainties we shall begin our analysis of the effects of rising energy prices on housing owners by adopting a worst-case approach. We assume that while builders of new housing can freely take advantage of opportunities for substituting other factors for energy, the amount of energy required to operate existing property remains fixed. In this way we can estimate the maximum possible effect of rising energy prices. Later we shall relax these rigid assumptions and use what evidence is available to develop more realistic estimates.

To look at the effects on property values it is necessary to consider the whole time path of energy prices and their effects on the incomes of owners. For this purpose we have constructed a simple simulation model to generate a stream of income that can be capitalized into a property value.

On the basis of the evidence we have presented we assume that the cost of providing housing services through new construction will determine the price of these services throughout the market. The owner of an existing house will, we assume, use the same quantities of energy, labor, water and sewer service, and other current inputs that he did before. As the prices of housing services increase because of the rising price of energy, his revenue will grow. But it will grow less rapidly than the cost of his fixed energy use. Because expenses grow more rapidly than revenues, his income will decline. The rate at which it declines will depend on the size of his energy bills in relation to his other operating costs.

Data on the relation between energy costs and total operating costs are difficult to obtain; in very few studies has the total pattern of revenues and expenses of existing residential properties been considered. Some useful information was gathered, however, as part of the Housing Assistance Supply Experiment. In table 2 the revenues received in 1973 by the average rental unit in the Green Bay, Wisconsin, metropolitan area are allocated among capital, energy, labor, and land. The figures indicate clearly that energy costs constitute a significant portion of the total. In that year, nearly a fifth of all the revenues generated by the average rental unit were used to pay energy bills.

Although these figures make an essential point, it is important to keep in mind the fact that they refer to a particular time and place and in one sense describe an extreme case. Winters in Green Bay, Wisconsin, are

Table 2. *Payments to Factors of Production, Regular Rental Properties, Green Bay, Wisconsin, 1973*
Dollars per unit a year

Factor of production	*Payment*	*Percent of total*
Capital[a]	990	56.2
Energy	319	18.1
Labor	200	11.4
Land	251	14.3
Total revenue	1,760	100.0

Source: Kevin Neels, *The Economics of Rental Housing*, Rand Report R-2775-HUD (Santa Monica, Calif.: Rand Corp., 1982), p. 43.
a. Includes repair expenditures.

among the most severe in the country. Heating requirements there are substantial, and as a result, energy accounts for a larger share of total costs than it would for the average rental unit in the United States. On the other hand, the figures shown in table 2 refer to the year 1973, before the initial rise in energy prices associated with the Arab oil embargo had taken place. The huge rise in energy prices since then has increased the share of energy costs in the total.[14] For that reason, the figures shown in table 2 are probably an understatement of the present importance of energy costs in Green Bay. The net result of these two facts is that the function of energy shown in table 2 can probably be regarded as roughly typical of what would be found in the middle and northern portions of the country. For the moment, let us consider this case.

In making a judgment about how much of an increase in the price of energy could be absorbed by the existing housing stock, we need to know the magnitude of the landlord's or homeowner's income from the property, recognizing that homeowners receive their income in the form of housing services rather than cash. How far can energy bills rise before the property owner will no longer be able to pay his bills and will have to abandon or demolish the building?

The payments to energy and labor shown in table 2 both represent payments to outside suppliers.[15] Much of the payment to land goes

14. Kevin Neels, *The Economics of Rental Housing*, Rand Report R-2775-HUD (Santa Monica: Rand Corp., 1982), p. 43.

15. Some of the payments to labor shown in table 2 represent compensation of the landlord for the time spent in managing and maintaining the property. Since the landlord's unpaid labor substitutes for the hiring of an employee, we regard these payments to landlords as the equivalent of payments to outsiders.

Table 3. *Disposition of Payments to Capital, Regular Rental Properties, Green Bay, Wisconsin, 1973*
Dollars per unit a year

Recipient	*Amount*	*Percent of total*
Landlords	299	30.1
Income from building	217	21.9
Compensation for unpaid labor	82	8.2
Banks	254	25.7
Tenants	57	5.8
Employees of landlords	11	1.1
Local government	203	20.5
Other suppliers	166	16.8
Total	990	100.0

Source: Calculations by the authors from data presented in Neels, *Economics of Rental Housing*, and Neels, "Revenue and Expense Accounts for Rental Properties," Rand Note N-1704 (Santa Monica, Rand Corp., 1982).

directly to the landlord to compensate him for owning the site. He could use this money to pay his energy bills; the payment to land represents a real opportunity cost, however. The land presumably does have an alternate use. The owner could demolish the old building and replace it with a new, more energy-efficient building. It is reasonable, therefore, to assume that the property owner would not allow the income he earned from the lot to fall below what he could earn by converting the site to another use.

Money for energy bills has to come out of the payment to the building. If factor ratios in an existing structure are truly fixed, the building has no alternate use. The payment to the building is part of the payment to capital. Table 3 shows the division of the payment to capital among the principal actors in the rental market.

Payments to outside suppliers cover the expense of repairs to the building. Payments to employees of the landlord cover the salaries of the repairmen who maintain it. The landlord's compensation for unpaid labor covers the time he spends on maintenance. Payments to tenants compensate them for repairs they have made and provide a return on the capital they supply, principally appliances. The remaining components of the payment to capital—landlord's income from the building, payments to banks, and payments to local government—make up the reserve available for payment of energy bills.

The landlord will continue to pay his energy bills until they have grown to the point at which his income from the building has gone to zero. He will then face an important choice. In order to be able to meet his mortgage payments, he will have to put his own outside income into the property. He may decide to do this, thinking that the mortgage will eventually by paid off and leave him with a positive income from the property. Or he may decide that prospects for the future are too dim to justify investing anything else in the property. He will then stop making mortgage payments and allow the bank to assume title. The bank will have no choice but to sell the building at a loss to a new owner. The new owner will have lower mortgage payments and, initially at least, a positive income from the building.

The component of payments to capital that goes to the local government is the property tax on the building. The property tax bill is theoretically proportional to the value of the property. As the value of the building declines, so should the amount paid to the local government in taxes. Lags in the process of setting assessments guarantee that this decline in taxes will not proceed smoothly. Nonetheless, as the value of the building declines toward zero, the amount paid in property taxes should decline toward zero as well.

When there is no money left for the landlord to take as income, for the bank to take in mortgage payments, or for local governments to take in taxes, the building becomes worthless.[16]

The bottom line is that the earnings of landlords and homeowners from the existing stock of housing are large enough to absorb a substantial increase in energy costs, but energy cannot be dismissed as inconsequential in relation to the earnings of a dwelling.

At first blush, the numbers in tables 2 and 3 might cause some alarm, since they indicate that if energy prices trebled the earnings of the existing stock of housing would be nearly wiped out. In such a view, however, two points are missed. First, part of any increase in energy prices will be passed on to the consumer as the cost of providing housing services rises for *both* new and old dwellings. Second, owners of existing dwellings will be able to take some steps to increase the energy efficiency of their structures, thereby escaping some increase in energy costs.

16. In actuality, as the value of the building fell toward zero, the landlord would probably begin to reduce the amounts spent for maintenance. This would provide some additional reserve. It is difficult to say how much maintenance would be reduced, however, or how rapidly the reductions would be made. In the interest of preserving our worst-case approach we have assumed that it is impossible for the landlord to divert funds from his maintenance budget in order to pay his energy bills.

Figure 1. *Effects of Real Energy Price Inflation of 2 Percent a Year, Assuming No Adjustments in the Existing Stock of Housing*

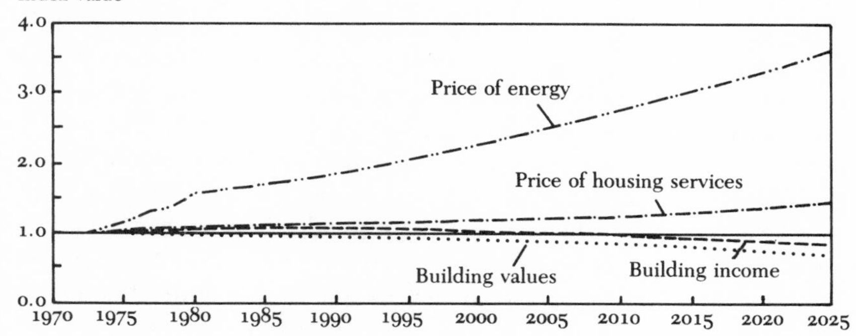

As an illustration of the first point, we have conducted several simulations, using the Neels production function for housing services. The simulations track four variables: the price of energy, the price of housing services, the owner's income from an existing dwelling, and the value of that dwelling.

For our first simulation, we shall assume that the prices of natural gas and heating oil each rise in real terms at an annual rate of 2 percent after 1980. At this rate, real energy prices will double in thirty-five years. In the case of oil, this assumption is in line with recent price forecasts.[17] The assumed rates of increase in gas prices are somewhat lower than current short-term and medium-term forecasts. In projections for natural gas it is generally assumed that with the removal of controls, prices will increase quite rapidly until they become comparable per unit of energy with prices for heating oil. Subsequent increases will be more modest. The inflation rate of 2 percent in real gas prices was chosen to reflect the average increases that could be expected during a period of several decades. Electricity prices are projected to rise in real terms after 1980 at an annual rate of 1 percent; for the period 1973–80 all energy prices follow their actual historical paths.

The results of this first analysis are shown in figure 1, in which are plotted indexes of energy prices, the prices of housing services, building incomes, and the values of existing residential buildings. All amounts are plotted in constant 1973 dollars. We abstracted from other trends in

17. See, for example, Foster Associates, "Fuel and Energy Price Forecasts," EPRI E A-411 (Washington, D.C., 1977).

housing markets, so the changes in prices and capital values should be interpreted as changes from what they would have been in the absence of increases in energy prices.

By the year 2025, energy prices had climbed to 360 percent of their constant-dollar 1973 values. The increase in the constant-dollar price of housing services was much more modest. By 2025, it was estimated to stand at roughly 142 percent of its 1973 level; an alternative way to express this would be to say that energy price increases would drive real rents of rental dwellings and imputed rents of owner-occupied dwellings to a level 42 percent higher than they would otherwise have been by 2025. The lower rate of inflation in the real price of housing services illustrates the ability of builders of new housing to substitute other inputs for energy and the limited share of energy in total factor costs.

The gap between the prices of energy and of housing services, when corrected for the cost share of energy, provides a measure of the pressure that rising energy prices place upon the owners of the existing stock of housing. This pressure is reflected in the value of the buildings. Value is computed as the net present value of the stream of building income generated under the definition outlined earlier.[18]

Value declines very slowly. By 1990, building values had fallen only about 5 percent. Even by 2025 they had fallen only to 70 percent of their constant-dollar 1973 values.

Increases in the rate of energy price inflation accelerate the obsolescence of the existing housing stock. To illustrate this fact, we present the results of a second simulation in figure 2. The assumptions underlying these curves are identical to those outlined earlier, except that the rates of increase in real energy prices after 1980 have been doubled. Real gas and oil prices rise at an annual rate of 4 percent, implying that they will double in about eighteen years. Real electricity prices rise at an annual rate of 2 percent. Under these conditions, the rise in energy prices is much steeper than before. By the end of the period shown, in 2025, energy prices stood in real terms at eight times their 1973 levels. As before, the rise in the price of housing services was much more gradual. By the end of the period shown, the real price of housing services had risen to a level about 90 percent higher than its 1973 level. The increase

18. We have used a discount rate of 4 percent. This figure is very close to the long-term difference between the nominal mortgage rate and the inflation rate. See Kevin Neels and C. Peter Rydell, "Measuring Capital's Contribution to Housing Services Production," Rand Paper P-6587 (Santa Monica: Rand Corp., 1981).

Figure 2. *Effects of Real Energy Price Inflation of 4 Percent a Year, Assuming No Adjustments in the Existing Stock of Housing*

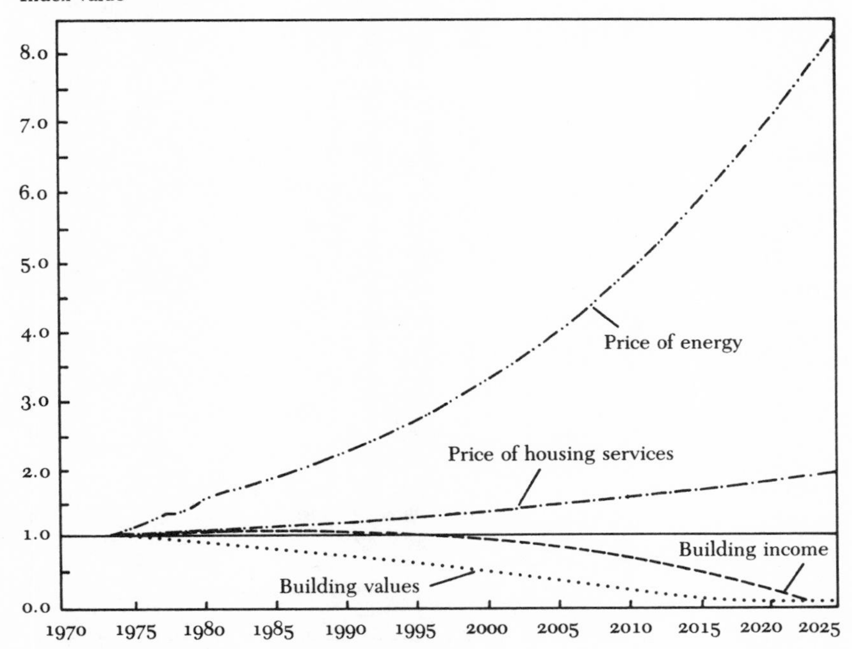

in the rate of energy price inflation caused a dramatic widening of the gap between the prices of energy and housing services, with a corresponding increase in the pressure placed upon the owners of the existing stock of housing. This pressure is reflected in a sharper drop in the value of existing residential structures. By 1990, building values were projected to fall to about 70 percent of their 1973 constant-dollar values. They were projected to fall to zero in the year 2022.

In both these simulations substantial reductions in the energy requirements of new dwellings are projected as the price of energy rises. By the year 2000, energy use in new dwellings is cut by a third in the first projection and by half in the second. When the value of the existing stock reaches zero, energy use in new dwellings is two thirds lower than in 1973 in both projections. These reductions in energy use, however, are derived from the use of the current technology by producers as captured in the Neels production function. To explore the possible effects of technological change during the coming years—as opposed to the

Figure 3. *Effects of Real Energy Price Inflation of 4 Percent a Year, Assuming No Adjustments in the Existing Stock of Housing and a One-Time Energy-Saving Technological Change in New Housing*

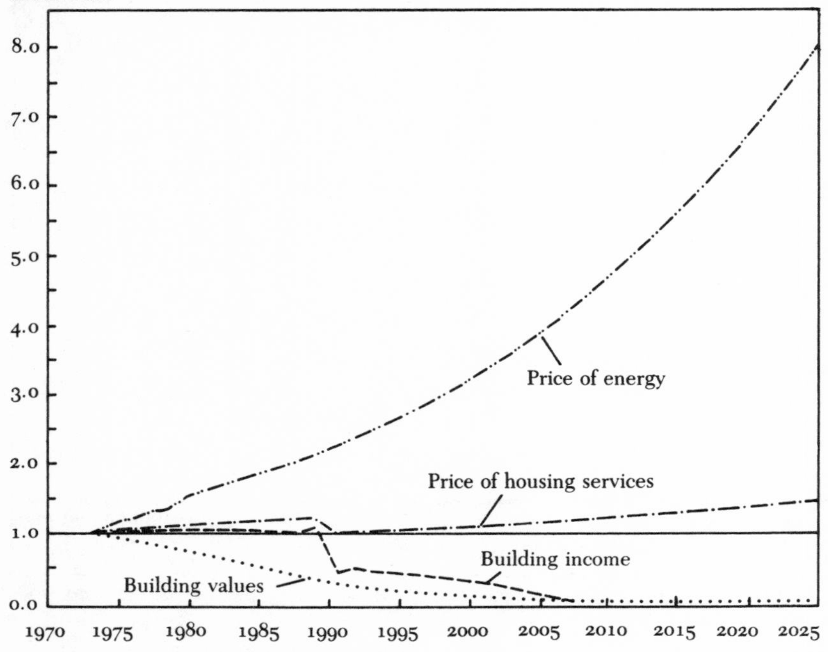

effects of substitution within the existing technology—we conducted a third simulation, in which genuine innovation was incorporated.

The possible effects of such an energy-saving technological change are illustrated in figure 3. To generate these curves, we assumed that in the year 1990, an energy-augmenting technological change will be introduced that allows owners of newly constructed housing to produce the same quantity of housing services using only half as much energy as before. The owners of existing housing will remain frozen at their original input levels.

The curve plotting energy prices in figure 3 is identical to that shown in figure 2. The curve of the price of housing services is significantly different. The introduction of the new technology causes a sharp drop in the price of housing services in 1991. The subsequent rise in the price of housing services is gradual. By the end of the period shown, in 2025, the prices of housing services have climbed to a level only about 50 percent higher than their initial real levels.

The effect of the new technology is to widen the gap between the prices of energy and housing services and to increase the pressure on the existing stock. We see this pressure reflected in the values of existing residential structures. Initially, they fall rapidly, dropping by 1990 to a level equal to 30 percent of their real 1973 level. Subsequent declines in values are somewhat more gradual. They were projected to fall to zero in 2007.

The declines in building values shown in figures 1–3 are substantial, but two important factors are likely to reduce the actual losses significantly: Owners of existing dwellings can modify their structures to improve their energy efficiency, and the average energy efficiency of the existing stock tends to rise in time because the least energy-efficient dwellings tend to be withdrawn from the stock of housing first.

To answer the question how possibilities for retrofitting can improve the prospects of the existing stock, we will use the results of a recent analysis of energy consumption and building improvement behavior in housing.[19] Rental properties were chosen to be studied here for several reasons. First, landlords and homeowners work with essentially the same technology and face the same possibilities for substituting other factors for energy. In this sense, analyses of the behavior of the two groups will yield similar results. Second, it is widely believed that landlords are less likely than homeowners to make energy-saving investments.[20] Examination of the behavior of landlords continues, in a sense, our worst-case analysis. Finally, there has been almost no research into the behavior of landlords or the energy efficiency of the rental housing stock.

These models describe the operating behavior of landlords and tenants during the period 1973–77 in Green Bay, Wisconsin, and South Bend, Indiana. Econometric estimates were made using data collected as part of the Housing Assistance Supply Experiment; these estimates provide a concise, empirical description of the way landlords reacted to the first substantial rise in energy prices.

We have probably understated the propensity of landlords to upgrade

19. Kevin Neels, *Energy Use in Housing*, Rand Report R-3030-DOE (Santa Monica: Rand Corp., 1983).

20. This argument is made by Richard M. Counihan and David Nemtzow in "Energy Conservation and the Rental Housing Market," *Solar Law Reporter*, vol. 2 (March–April 1981), pp. 1103–32.

their buildings. At the time of the original Arab oil embargo it was widely believed that the resultant increases in energy prices would be temporary. We can therefore assume that landlords responded less to rising energy prices at that time than they would now. In that sense, even our best-guess estimates of the effects of energy price inflation err in the direction of understating the viability of the existing stock.

We have taken the inflexibility of the existing stock of housing into account with another model developed in the Neels study, in which the way the quantity of capital on an existing property changes in response to variations in the maintenance and improvement policies of landlords is described.

Detailed descriptions of all these models can be found in the source document for the Neels study.[21]

The simulation depicts the behavior of a particular landlord who owns a particular rental property. The model was run for the period 1973–2025. For the period 1973–80 we used actual prices of energy, labor, and capital. For the period from 1981 onward we assumed that the prices of labor and capital would remain constant in real terms while energy prices increased at the same rates that we used in the foregoing analyses. The parameters of the model were set so that it would describe the same rental property that we examined earlier.

The outcome generated by the model represents our best estimate of what can be expected to happen to the existing housing stock during the coming decades. The behavior predicted by the model is much richer than that which we assumed in generating our worst-case projections.

Figure 4 shows what happens when real energy price inflation is 2 percent a year and landlords are permitted to upgrade their buildings. The curves for the real prices of energy and housing services are identical to those shown in figure 1. The curves for income realized from buildings and the values of the buildings reflect the effects of the upgrading efforts of landlords.

The decline in current income shown here is somewhat more rapid than that shown in the worst-case analysis. It reflects the diversion of some of the income of landlords into efforts to improve their buildings. These efforts bear fruit in the form of more stable capital values. By the year 2025 in the worst-case analysis building values had fallen to levels less than 70 percent as high as their initial levels and were declining

21. Neels, *Energy Use in Housing.*

Figure 4. *Effects of Real Energy Price Inflation of 2 Percent a Year, Assuming Upgrading of the Existing Stock of Housing*

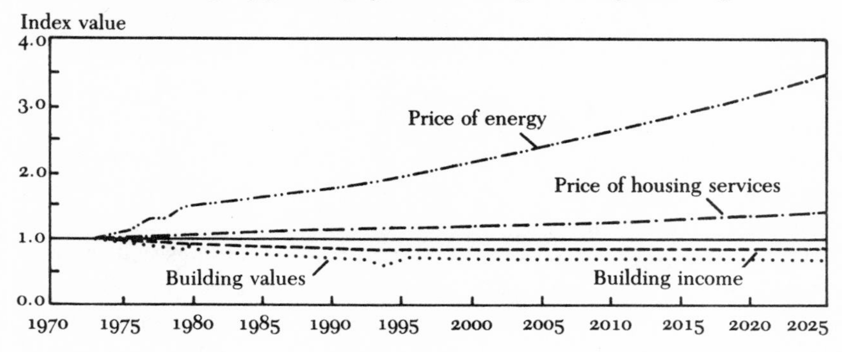

Figure 5. *Effects of Real Energy Price Inflation of 4 Percent a Year, Assuming Upgrading of the Existing Stock of Housing*

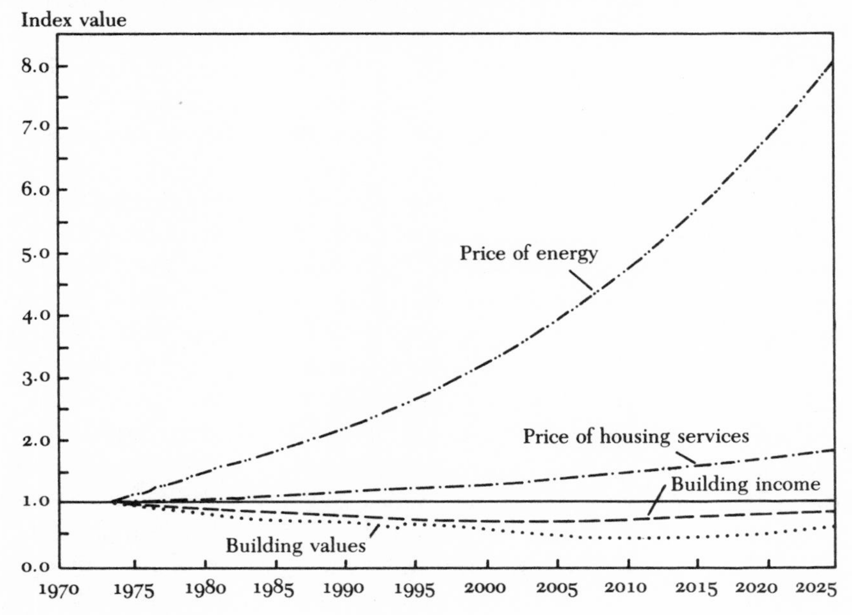

rapidly. In contrast, when landlords were allowed to improve the efficiency of their buildings, values had, by 2025, stabilized at levels a little higher than 80 percent of their initial level.

Figure 5 shows what happens when landlords are confronted with 4 percent inflation in real energy prices a year. The real prices of energy and housing services follow the same paths shown in figure 2. The patterns of incomes from buildings and their values is more complex, however. Initially incomes decline rapidly; during the second decade of the next century they reach bottom at a bit less than half their initial levels. Then they begin to climb. The pattern of building values is similar, though more muted. The value of a building reaches the bottom shortly after the turn of the century at about three quarters of its initial level, then begins a slow climb.

We are not ready to predict that all apartment buildings will share in this rosy future. The landlord whose behavior we have depicted faces a steady demand for his product by tenants who have sufficient income to pay the rent. If the demand for rental housing falls, either because of a shift of households to homeownership or because of a decline in the real incomes of renters, landlords will have much less of an incentive to improve the energy efficiency of their buildings. What the simulation does show is that there are possibilities in the technology for reducing energy use and that landlords are willing to take advantage of them. The simulation shows, moreover, that the necessary improvements can be financed out of the income of the property without massive infusions of outside capital.

The simulation also shows that while it is eminently in the interest of landlords to invest in energy-saving improvements, it may not be in their best interest to do so immediately. This is a point often overlooked in studies of the efficacy of energy-saving innovations. The rate of return from a particular project may be high enough to justify undertaking it today if the alternative is never to undertake it, but the return might be still higher if it is postponed until tomorrow.

Natural gas, for example, the source of most of the energy used in the residential sector, is even now relatively inexpensive. At the outset of the simulation the amount of money that could be saved by improving energy efficiency is not sufficient to offset the cost. The landlord therefore lets the building begin to run down while energy use increases. The real upgrading effort begins much later, after energy has become considerably more expensive. Real savings in energy do not begin to be seen until the middle of the next decade.

Even with this slow start, the capital losses suffered by owners appear to be fairly modest. At their lowest points real building values fall to levels only 20 to 25 percent lower than their initial levels. The same point can be made here that was made earlier in connection with housing consumers. These losses, if realized, will be substantial, but they will hardly constitute a crisis.

It is likely that the actual capital losses suffered will be smaller than those shown here. The second factor that will tend to increase the long-term viability of the existing stock is that the most energy-efficient dwellings are the ones that will tend to remain longest in the stock. To illustrate this point, we shall again draw from data gathered in the Housing Allowance Supply Experiment.

The average amount of energy used by all rental units in Green Bay, Wisconsin, and South Bend, Indiana, was compared to the amounts used by units that were removed from the rental stock during a three-year period. The units that were eventually retired used 16–28 percent more energy than the average rental unit. In South Bend the removal of dwellings that used large quantities of energy accounted for most of the reduction in average energy use there in the years immediately following the Arab oil embargo. Hence, although higher energy prices may hasten the withdrawal of some energy-inefficient dwellings from the market, the remaining stock of dwellings will be better able to compete with newly constructed housing.

In sum, the principal message that emerges from our simulations is that energy prices would have to rise considerably more than they are now expected to rise before the existing stock of housing is in any danger of becoming economically unviable. Even with no technical adaptation by the owners of existing dwellings, according to our most likely projection of energy price inflation, we estimate that it will be well into the next century before energy bills exhaust the incomes of the owners. The introduction of a highly significant energy-saving technological change in an environment of massive energy price increases shortens this grace period only to thirty years. It is clear, therefore, that there is no immediate crisis.

The ability of property owners to improve the energy efficiency of their properties and their tendency to remove the least efficient dwellings from the stock will do much to mitigate these adverse effects. While rising energy prices will impose burdens on the owners of the existing housing stock, just as they will impose burdens on the rest of society, these will be burdens that the owners can bear.

Conclusions and Policy Implications

Our review of engineering studies of residential energy use has shown that the technology available today for the production of housing services presents ample opportunities for reducing the consumption of energy in housing. These opportunities, however, are not uniformly available to owners of housing. The builder of new housing can alter his design and freely substitute other inputs for energy. In contrast, the owners of existing housing are far more constricted. Many of their options were foreclosed at the time of construction, and many of the opportunities open to them now come at a much higher cost.

This divergence between the two sectors can place severe cost pressure on the owners of the existing stock of housing. Competitive pressures guarantee that the prices of housing services will rise by no more than the amounts that builders of new housing must charge to cover their costs. Their costs will be low because they will have taken steps to reduce their energy requirements. Owners of existing housing will not be able to realize the same gains in efficiency. Their revenues will reflect the lower costs of new housing. Their higher costs will have to be paid out of their income from the building. If the divergence in costs between new and existing housing becomes large enough, the income of existing properties could fall to zero and they might be abandoned.

We find little reason for alarm over this possibility, however. Even in the unlikely event that owners of existing buildings do nothing to reduce their use of energy, owners of new housing will pass along enough of the energy price increase, and energy costs will be low enough in relation to income, that we will be well into the next century before any danger that the existing stock will become economically unviable will arise. Even if a potentially serious problem exists, there is ample time in which to think of ways to deal with it.

Our analyses indicate that owners of existing housing will deal with the problem effectively. Rising energy prices will accelerate the retirement of the most inefficient members of the housing stock. The buildings that remain will be modified in ways that will reduce their energy requirements. The technology now available is flexible enough that even existing buildings can make a smooth transition to a regime of higher energy prices. Our examination of actual behavior of landlords, moreover, indicates that they are perfectly capable of responding adequately to the incentives presented by rising energy prices. We expect that owner-occupants will, if anything, be even more ready to upgrade their dwellings. If owners do not now appear to be doing enough to reduce energy use, the apparent reason is that energy prices have not yet risen high enough to make the required changes worth while.

We conclude that whatever problems are being created for the existing housing stock by rising energy prices are well on their way to being taken care of. This suggests that the most appropriate public policy would be passive. Subsidies and mandatory energy-efficiency standards may not be needed. Instead the government should aid the private market in making its own adjustment.

The government should take steps to make sure that the prices paid by consumers for energy accurately reflect the full costs to society. This means a gradual removal of price controls and subsidies and a move from average-cost to marginal-cost pricing by utilities. The studies we have drawn upon indicate that property owners do respond to energy prices. Holding prices at artificially low levels encourages waste and delays much-needed adjustments to the new energy supply situation.

The government should also take steps to make sure that all parties have access to enough information to make informed decisions about energy. Public-utility commissions should encourage the use of report-card billing so that consumers will have a clear idea of the prices they are paying, the amount of energy they are using, and how their use compares with that of other consumers. State and federal governments should disseminate information about the costs and energy savings associated with various improvements in the existing stock of housing. Utility bills should be made matters of public record so that prospective buyers and tenants can judge the probable energy costs of housing units.

Giving owners the information they need will be useless, however, if imperfections in the capital market prevent their acting on it. Special loan programs for low-income homeowners may be necessary if these homeowners are to finance economical innovations. Furthermore, the government should find out why long-term financing of energy-saving investments is often unavailable to small landlords.

Another frequent concern of policymakers is that increases in energy prices may place an extreme burden on the poor and the elderly, who tend to live in the least energy-efficient dwellings. Our analysis suggests that such a concern for poor or elderly homeowners might be well placed, but that with respect to renters—and most of the poor do rent—the apprehension is probably unwarranted. The evidence presented here indicates that it is the owners, not the tenants, of the least efficient dwellings who are most likely to bear the brunt of higher energy costs.

In sum, we find little evidence that rising energy prices have yet precipitated or are likely to precipitate a crisis in the existing housing stock. Even as it now stands, the private market appears to be doing reasonably well in dealing with the problem. And while governments at all levels could do much to aid this adjustment process, there appears to be little need for massive, direct public intervention.

Appendix: Translog Production-Function and Cost-Function Models

The analysis reported in the text relies upon the description of the technology of producing housing services contained in the coefficients of a translog production function and a translog cost function. The coefficients of the production function were derived from data collected in the Housing Assistance Supply Experiment.[22] The model was estimated directly; that is, the dependent variable for the regression was the output of housing services, and the explanatory variables were transformations of the input levels of capital, energy, labor, and land. All inputs were measured in physical units. The quantity of capital was measured by an index constructed from measures of physical attributes, such as the number of rooms, the stock of appliances, and the condition of the structure.[23] Energy was measured in millions of British thermal units (Btu) of energy a month. Labor was measured by what a dollar would buy in 1973. Land was measured in square feet.

The coefficients of the production function are shown in table 4.

The cost function was derived from the coefficients of the production function. It represents an approximation of the true cost function associated with the production function.

To derive the coefficients of the cost function we first computed the Allen partial elasticities of substitution using the coefficients of the production function. We then used the formulas that relate the partial

22. For a more detailed description of the production function and its estimation, see Neels, "Derived Demand for Energy in the Production of Housing Services."

23. This index is described by Neels and Rydell in "Measuring Capital's Contribution to Housing Services Production."

Table 4. *Translog Production-Function Coefficients*

Term[a]	*Coefficient*	*Standard error*
K	0.29145	0.28258
E	0.16548	0.13010
N	0.18636	0.10497
L	0.36438	0.13248
*K***K*	0.03562	0.02119
*K***E*	−0.04551	0.01681
*K***N*	−0.01900	0.01328
*K***L*	−0.03292	0.01560
*E***E*	0.01955	0.00143
*E***N*	0.00623	0.00573
*E***L*	0.02831	0.00677
*N***N*	−0.00306	0.00332
*N***L*	0.00240	0.00574
*L***L*	−0.00936	0.00376
Constant	2.31753	[b]

a. *K* = log of quantity of capital
E = log of quantity of energy
N = log of quantity of labor
L = log of quantity of land.
b. The constant term was set to establish units of measure.

derivatives of the cost function to the Allen partial elasticity values to compute all the second-order cross-product coefficients for the translog. The remaining coefficients were derived using the cost-share equations and the cross-coefficient constraints that guaranteed that the function would be homogeneous of degree one.

Factor prices were all computed by dividing total expenditures for each factor of production by the measure of the physical quantity.

The coefficients that were computed in this way are shown in table 5.

Comment by Terry H. Morlan

In this paper the authors have addressed the question whether rising energy prices will adversely affect the viability of the existing stock of housing. After establishing evidence that significant reductions in energy intensity are feasible with present technology, the authors examine the theoretical aspects of the issue and attempt to quantify the possible effects of rising energy prices on the costs of housing services and on the value of the existing stock of houses. Their conclusion is that rising energy

Table 5. *Translog Cost-Function Coefficients*

Term[a]	Coefficient
PK	−0.49555
PE	−0.24322
PN	0.35924
PL	1.37953
PK*PK	−0.50560
PK*PE	0.23795
PK*PN	0.09252
PK*PL	0.17513
PE*PE	0.05302
PE*PN	−0.09293
PE*PL	−0.19804
PN*PN	0.07753
PN*PL	−0.07712
PL*PL	0.10003
Constant	2.09708

a. *PK* = log of price of capital
PE = log of price of energy
PN = log of price of labor
PL = log of price of land.

prices are likely to increase the cost of housing services and impose capital losses on owners of existing housing. The authors conclude further, however, that these effects will not be large and will not jeopardize the viability of the existing housing stock until well into the twenty-first century.

The examination of technical possibilities for reducing energy requirements in buildings indicates that there are substantial possibilities from such investments as insulation, storm doors and windows, and furnace modifications. The key conclusion for the rest of the analysis is that the possibilities for substituting capital for energy in providing housing services are significantly greater for new structures than for existing structures. This disparity forms the basis of the authors' conclusion that rising energy prices will affect the value of the existing stock of housing adversely. I think that the analysis is weakened somewhat by the fact that the only housing market adjustment mechanism considered is the substitution of capital for energy.

The first conclusion, that increasing energy prices will increase the costs of housing services, is not surprising. There seems to be some agreement among studies of residential energy price elasticity that the elasticity is negative and less than unity in absolute value. If this is true, an increase in price would be accompanied by an increase in expenditures. Neels and Murray have presented evidence that, because there are some opportunities for conservation, even in existing structures, and because energy is not the dominant factor in the cost of housing services, the effect on housing markets should not be large. The problem that housing energy costs may constitute a significant share of the budget of a poor household is not addressed.

The second conclusion of the paper, that the value of existing housing will decline with rising energy prices, is less clear than the first. During the 1970s, a period of rapidly rising energy prices, the consumer price index for existing houses rose faster than the index for new houses. Although this does not appear to support the authors' conclusion, it does not necessarily contradict it. During the 1970s other things besides energy prices were unequal. According to data from the Federal Housing Administration (FHA) presented in Dwight Jaffee's paper, debt service and hazard insurance costs, which together accounted for nearly three quarters of the costs of homeownership in 1980, increased more than energy costs between 1974 and 1980. This fact would presumably favor the existing housing market and could easily swamp the effects of higher energy costs.

Nevertheless, some theoretical questions remain. As mentioned earlier, Neels and Murray appear to assume in the analysis that all adjustments to rising energy costs are made through factor substitution in the production of housing services. The advantage enjoyed by new housing in making such substitutions puts existing stocks at a disadvantage and causes them to decline in value.

The analysis could be expanded, however, by exploring further the implications of the authors' first conclusion, namely, that increased energy prices lead to higher prices of housing services. If the real prices of housing services should increase, consumers would probably substitute other expenditures for housing services—that is, the demand for housing services would probably decline. Consumers might accept a lower degree of comfort that is directly related to energy use—by turning down thermostats in winter, for example. It might also mean reducing other aspects of housing satisfaction, however, in such a way as to increase the demand for older houses or for multifamily housing, which requires significantly less capital, land, and energy per housing unit.

If the demand for existing housing increased as a result of higher energy prices, the loss in value predicted in the paper might be negated.

In the absence of other effects, the reduced housing-services-demand argument indicates that the multifamily housing share would grow during periods of increasing energy prices. Between 1970 and 1977 the Annual Housing Survey shows an increase in the multifamily share of occupied housing units from 30.3 percent to 30.9 percent. Data on housing starts and completions show significant increases in the multifamily share in every year since 1977. It is not clear, however, that these increases are the result of energy prices, since the changes are consistent with the expected effects of other increases in housing costs during the 1970s.

Two other ways in which energy prices might affect the existing stock of housing were not discussed by Neels and Murray. Since they are the subject of other papers in this conference, however, I will mention them only briefly. First, if higher energy prices were to affect urban configurations through higher transportation costs, the viability of the existing housing stock would also be affected. Second, if rising energy prices were to affect regional industrial location and population shifts, they would also affect the value of existing housing in areas that were losing population.

The last section of the paper purports to deal with "actual energy-conserving behavior of landlords." I found this section weak in that "actual" behavior turned out to be a model simulation. It requires a great deal of faith to accept a model simulation as representative of actual behavior. Unfortunately, not enough information about the model is given to make clear what kind of behavior is assumed in the model.

Some information on actual energy conservation activities, which should give a better indication of actual behavior, is now available from the Energy Information Administration (EIA). It should also give some indication of long-term responses to differences in costs of energy-related housing services in that it gives information on variations in housing characteristics across regions.

It is also worth noting that investigations of the effects of energy prices on urban configuration and regional location patterns indicate that these effects are small. Neels and Murray found them to be small, and my reservation would tend to indicate that they are smaller still. Taken together the papers from this conference seem to offer an optimistic view of the ability of the U.S. economy to adjust to higher energy prices during the 1980s. In the long run, actual shortages of conventional fuels may be more difficult to cope with. Conservation and a smooth transition to alternative sources of energy therefore continue to be important.

DWIGHT M. JAFFEE

New Residential Construction and Energy Costs

IN this paper I shall survey the effects of high and rising energy prices on new residential construction. For the purposes of this analysis, it is assumed that the real price of energy for residential use will not decline significantly from the current peak, and that it may well rise further. The response of new housing construction to this level of energy prices can be analyzed in aggregate flows of new construction; characteristics of newly constructed units; differential effects by type of structure and by region; the trade-off with existing units; other aspects of the supply of housing—the construction industry—and the demand for housing—the typical buyer; and the importance of housing finance.

The starting point of the analysis is the assumption of high and possibly rising real energy prices. These high energy prices obviously lead to higher costs for residential energy if energy-conservation features remain unchanged in new units. And, in the short run, before new technology and standards can be introduced, higher costs for residential energy and use of less energy—that is, a lower level of housing service—are the primary effects of higher energy prices.

After a lag, new technology and new construction techniques are employed and new standards are set to offset the primary effects of the higher energy prices. At one extreme, it is possible to imagine an energy-conserving technology so efficient that residential energy costs would remain unchanged—without reducing the level of housing services—even in the face of higher energy prices. Putting new technology to use, however, entails significant capital costs—that is, construction costs are higher. In particular, among the cost-efficient technologies there will be a trade-off between the conservation of energy—and concomitant savings in cost—and the capital costs of putting the technology to use.

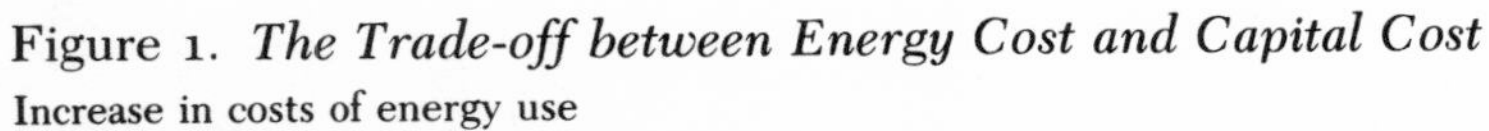

Figure 1. *The Trade-off between Energy Cost and Capital Cost*

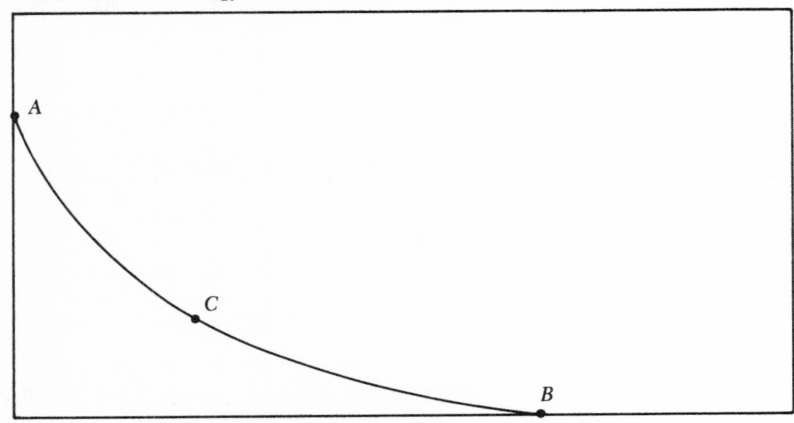

The trade-off between higher energy costs and conservation-related capital costs is also illustrated in figure 1 by the curve *AB*. Point *A* corresponds to the situation in which no new energy-conservation technology is introduced, and energy costs rise accordingly. Point *B* is the opposite extreme, just discussed, in which the energy-conserving technology introduced is so efficient that energy costs actually remain unchanged. Point *C* represents a more typical situation, in which the increase in the cost of energy used is maintained at an intermediate level through energy-conserving capital expenditures.

The right amount of capital investment in energy-saving technology for the trade-off depends on a variety of factors. In basic theory, the efficient point is reached where the saving in the cost of using additional energy is just offset by the additional capital costs that are necessary to achieve the saving. But in practical terms a number of factors complicate the determination of this point. First, the curve of figure 1 pertains to the *efficient* technology, where efficiency is defined as providing a given saving in the cost of the energy used at the lowest possible capital costs. The efficient trade-off curve, however, requires full information on existing technology, and the curve will shift as new technology becomes economically efficient through innovation or economies of scale. Second, since the capital expenditure is an immediate expense while the savings in energy use accrue gradually, there is a time element involved in the decision that requires the use of the correct interest rate. Third, energy-efficiency standards may be enforced at levels higher than the private-market participants would otherwise set. The decision on this trade-off is not made only at the technological level. For the most part, it can be assumed instead that the decision is being made effectively, with the capital costs of the investment and the requirements imposed by energy-efficiency standards taken into account. My focus, then, is on the way this decision affects new housing construction.

The Effects of High Energy Prices on Housing Construction: General Considerations

In December 1980 the Economics Division of the National Association of Home Builders (NAHB) tabulated the results of an extensive survey of the attitudes, preferences, interests, and outlook of new home buyers during 1977 and 1978. A number of the questions had to do with energy, and the material is an excellent source of information concerning consumers' perceptions of the effects of energy prices on housing. The sample was generated through the NAHB Home Owners Warranty (HOW) Program, in which participating builders completed enrollment forms for each unit. The characteristics of the HOW homes are very much like the characteristics of new homes as developed by the Bureau of the Census. Thus, although the sample is of modest size—5,000 questionnaires mailed, about 1,400 useful responses—it is valuable for the purposes of this study.

The responses to the relevant energy-related questions of the survey are tabulated in table 1, which includes the results both for the country as a whole and disaggregated according to the nine regions of the U.S. Census.

Table 1 provides the responses to four questions concerning consumers' general perceptions of the energy situation. Consumers were asked to rate the energy situation five years earlier—that is, in about 1973. Only about 7 percent of the respondents rated the situation as extremely serious. They were then asked to rate the energy situation "now"—that is, in 1978. More than 36 percent of the respondents rated the energy situation in 1978 as extremely serious. The third question asked of the respondents was to rate the energy situation "five years from now"—that is, in about 1982. More than 74 percent of the respondents saw the

Table 1. *Consumers' Perceptions of the Energy Situation, by Census Region*
Percent of respondents

Rating of the energy situation	*New England*[a]	*Middle Atlantic*[b]	*East North Central*[c]	*West North Central*[d]	*South Atlantic*[e]	*East South Central*[f]	*West South Central*[g]	*Mountain*[h]	*Pacific*[i]	*Total*
The situation five years ago										
Extremely serious	8.8	7.9	8.2	3.7	6.4	6.7	10.1	5.6	3.9	7.0
Serious	21.9	21.3	14.9	19.5	19.1	14.4	20.2	13.9	21.1	18.4
Somewhat serious	36.8	31.5	37.3	40.2	42.7	37.5	30.3	37.0	38.2	37.2
Not really a problem	28.1	37.0	38.1	32.9	28.6	38.5	35.8	42.6	31.6	34.4
Not sure *or* do not know	4.4	2.4	1.5	3.7	3.2	2.9	3.7	0.9	5.3	3.0
Total	100.0	100.0	100.0	100.0	100.0	100.0	100.0	100.0	100.0	100.0
Number of respondents	114	127	134	82	220	104	109	108	76	1,074
The present situation										
Extremely serious	33.0	38.6	33.8	37.9	38.7	35.2	38.7	29.8	30.0	36.1
Serious	51.5	44.7	48.3	44.8	48.8	41.9	46.2	56.1	44.3	47.7
Somewhat serious	13.4	14.4	14.5	13.8	9.7	21.0	14.2	10.5	15.2	13.6
Not really a problem	1.0	1.5	2.8	2.3	2.3	1.0	0.9	2.6	1.3	1.8
Not sure *or* do not know	1.0	0.8	0.7	1.1	0.5	1.0	0.0	0.9	1.3	0.7
Total	100.0	100.0	100.0	100.0	100.0	100.0	100.0	100.0	100.0	100.0
Number of respondents	97	132	145	87	217	105	106	114	79	1,082

The situation five years hence										
Extremely serious	72.6	76.2	70.1	79.5	75.4	61.3	75.4	84.7	72.1	74.2
Serious	9.7	17.5	19.5	7.7	13.9	25.8	18.0	12.5	20.9	16.1
Somewhat serious	4.8	4.8	3.9	7.7	5.7	3.2	0.0	1.4	0.0	3.7
Not really a problem	4.8	0.0	2.6	2.6	2.5	1.6	0.0	0.0	0.0	1.7
Not sure *or* do not know	8.1	1.6	3.9	2.6	2.5	8.1	6.6	1.4	7.0	4.3
Total	100.0	100.0	100.0	100.0	100.0	100.0	100.0	100.0	100.0	100.0
Number of respondents	62	63	77	39	122	62	61	72	43	601
The situation in the future[j]										
No shortage	5.0	5.5	3.7	2.1	3.2	8.2	4.3	1.6	6.7	4.3
Shortage likely	67.5	59.3	59.5	67.0	67.6	63.9	64.1	72.4	72.2	65.6
Private industry will solve the problem	40.8	31.7	31.3	31.9	35.6	27.9	47.9	37.8	32.2	35.2
Number of respondents	120	145	163	94	253	122	117	127	90	1,231

Source: National Association of Home Builders, Economics Division, "Decisions for the '80s" (Washington, D.C.: NAHB, 1980), pp. 60–63. Figures are rounded.

a. Maine, New Hampshire, Vermont, Massachusetts, Rhode Island, and Connecticut.

b. New York, New Jersey, and Pennsylvania.

c. Michigan, Ohio, Indiana, Illinois, and Wyoming.

d. Minnesota, Iowa, Missouri, Kansas, Nebraska, South Dakota, and North Dakota.

e. Delaware, Maryland, West Virginia, District of Columbia, Virginia, North Carolina, South Carolina, Georgia, Florida, and Puerto Rico.

f. Kentucky, Tennessee, Mississippi, and Alabama.

g. Arkansas, Louisiana, Texas, and Oklahoma.

h. Montana, Wyoming, Colorado, New Mexico, Arizona, Utah, Idaho, and Nevada.

i. Washington, Oregon, California, Hawaii, and Alaska.

j. Responses in five other categories, which had to do with government intervention to deal with the energy situation, were also available.

situation as extremely serious. Finally, respondents were asked to characterize the energy situation of the future. More than 65 percent indicated that they thought a shortage of energy likely. Overall, it seems that between a third and three quarters of the respondents considered the energy situation between 1978 and 1982 to be extremely serious, including a likely shortage of energy. They perceived that the situation was deteriorating during the 1970s and expected this trend to continue.

Table 2 concerns those characteristics of housing units that consumers deemed the most important when making a new purchase. The first question concerned the most important considerations in buying a home. Almost 80 percent of the respondents said that they would like a more energy-efficient home. A bigger house was the next most common response, but at the much lower level of about 43 percent.

Respondents were then asked about features that were important when a home was last purchased or that would be important in the next home purchased. As in the previous question, energy efficiency was one of the choices, but the alternatives in these questions were different. Energy efficiency was again a common response, indicated in about 66 percent of the responses regarding the present home and about 80 percent of the responses regarding the next home. Features such as the reputation of the builder, the neighborhood, and high-quality workmanship, however, were even more common responses.

The fourth question was whether the new home was advertised as energy-efficient. The responses were about evenly split, nearly 46 percent yes, nearly 43 percent no, and about 11 percent unsure.

The fifth question was about the type of structural units preferred. Nearly 93 percent of the respondents favored single-family, detached units; there were only scattered responses in favor of the other types. Since only about 70 percent of the present U.S. stock of housing is single-family units, these responses represent an overwhelming preference for the single-family unit. It should be noted, however, that the survey sample included only households already living in single-family units.

Table 3 concerns the conservation of energy. The first questions asked were whether consumers would be willing to pay $500 and $1,200–1,500, respectively, for extra insulation. About 62 percent were quite willing, while about 7 percent were not at all willing, to spend $500; about 45 percent were quite willing, and only about 16 percent were not at all willing, to spend $1,200–1,500.

The last three questions concerned the willingness of consumers to alter their heating systems. Responses to the third question indicated that about 19 percent of the consumers were considering the installation of another type of heating system. Responses to the next question indicated that of those altering their systems, the large majority—about 82 percent—were adding supplemental heating, while the remaining 18 percent were converting the primary system. The fifth question listed the types of heating being considered. Wood burners were the most popular choice by a large margin, followed by electric heat pumps and solar heating systems.

The data in tables 1–3 are shown for the nine U.S. Census regions as well as for the United States as a whole. Some tendencies manifested themselves in the regional data. In particular, responses to a number of the questions made the Northeast, the South, and the Mountain region all appear more sensitive to energy considerations than the national average. Even these responses, however, are sporadic; within the Northeast, for example, sometimes New England appears more sensitive and sometimes the Middle Atlantic region appears more sensitive. Thus, while substance could be read into the regional variations in the responses, the relatively small size of the sample probably limits their value.

Returning to the national trends, it appears that from a third to three quarters of the households surveyed were quite sensitive to the importance of energy considerations in their housing choices. This includes characterization of the energy situation as extremely serious, indications that energy efficiency was among the most important considerations and features in home-purchase decisions, and action on a variety of conservation alternatives.

Conservation of Energy

The actions taken by consumers with regard to residential energy conservation show another facet of the importance attached to the effects of rising energy prices. One aspect of these actions was shown in the large number of households that were found to be considering supplemental heating systems. Although energy-conservation activities affect only older units directly, they do suggest the characteristics desired by consumers in newly produced units.

As part of the National Interim Energy Consumption Survey, the U.S. Department of Energy carried out two surveys having to do with residential energy consumption during the winter of 1978–79. One survey

Table 2. *What Consumers Seek in the Purchase of a Home, by Census Region*[a]
Percent of respondents

Question and response	*New England*	*Middle Atlantic*	*East North Central*	*West North Central*	*South Atlantic*	*East South Central*	*West South Central*	*Mountain*	*Pacific*	*Total*
If you were going to move, what would be the most important considerations in your selection of a home?										
Greater energy efficiency	78.6	87.7	75.0	70.8	80.3	85.6	81.4	81.4	68.2	79.5
Larger lot	33.3	32.2	48.2	42.7	36.5	34.7	44.9	41.1	50.0	39.8
Larger house	35.9	30.8	47.6	47.2	40.6	48.3	55.9	40.3	42.0	42.7
Better neighborhood	23.9	19.9	19.5	19.1	26.9	28.0	24.6	33.3	31.8	25.1
New house	36.8	37.0	28.7	47.2	40.6	33.1	28.8	29.5	30.7	34.9
Smaller house	10.3	13.0	7.3	6.7	12.4	7.6	7.6	7.8	14.8	9.9
Smaller lot	3.4	8.2	1.8	2.2	6.8	4.2	3.4	5.4	2.3	4.6
Old home to rehabilitate	5.1	2.7	5.5	3.4	4.4	7.6	6.8	6.2	5.7	5.2
Older home	1.7	3.4	3.0	4.5	2.0	1.7	4.2	7.0	1.1	3.1
More luxury features	22.2	28.8	29.3	37.1	29.3	31.4	38.1	38.0	30.7	31.2
Low-tax neighborhood	26.5	35.6	12.8	19.1	20.9	18.6	20.3	16.3	9.1	20.4
Condominium or cooperative	6.0	2.1	3.0	4.5	3.6	5.1	3.4	4.7	1.1	3.7
Town house	1.7	5.5	1.2	3.4	3.6	3.4	6.8	2.3	2.3	3.4
Mobile home	0.0	0.0	0.6	0.0	0.8	0.0	0.8	0.8	1.1	0.5
Inner-city location	2.6	2.1	0.6	2.2	1.6	5.1	0.8	1.6	4.5	2.1
Other	6.8	7.5	11.0	3.4	7.2	8.5	7.6	10.1	11.4	8.2
Number of respondents	117	146	164	89	249	118	118	129	88	1,218
What features were important to you when you bought a new home?										
Reputation of builder	73.6	66.7	69.9	64.1	68.4	58.5	73.0	58.5	58.4	66.3
Neighborhood	91.7	84.7	90.4	85.9	91.0	92.7	91.0	83.1	89.9	89.1
Quality of workmanship	81.0	74.7	79.5	77.2	79.7	74.0	79.5	76.9	77.5	78.0
Warranty	69.4	67.3	69.3	68.5	79.7	66.7	72.1	65.4	73.0	71.0
Proximity to shopping area	32.2	29.3	39.8	33.7	48.4	39.0	33.6	33.8	43.8	38.1
Proximity to work	61.2	48.0	48.2	47.8	48.4	51.2	46.7	53.8	57.3	50.8
School district	43.8	44.7	55.4	54.3	42.6	48.0	49.2	39.2	52.8	47.1
Energy-saving features	67.8	54.7	55.4	56.5	63.3	60.2	67.2	62.3	51.7	60.3
Recommendation of another buyer	7.4	7.3	7.8	6.5	8.2	8.1	5.7	9.2	7.9	7.7
Other	9.9	13.3	12.7	7.6	13.3	12.2	13.9	8.5	11.2	11.8
Number of respondents	121	150	166	92	256	123	122	130	89	1,249

What features would be important to you if you were buying another new home?										
Reputation of builder	86.4	84.1	83.0	84.1	78.0	78.5	81.3	74.3	67.9	79.9
Neighborhood	86.4	79.5	83.0	81.8	78.9	82.2	77.6	74.3	78.2	80.1
Quality of workmanship	87.4	87.1	85.7	83.0	83.6	87.9	86.9	83.2	84.6	85.4
Warranty	65.0	63.6	70.1	65.9	67.2	68.2	72.0	61.1	62.8	66.5
Proximity to shopping area	33.0	39.4	44.9	52.3	50.9	48.6	43.9	38.1	50.0	44.9
Proximity to work	64.1	56.8	65.3	62.5	56.0	57.9	56.1	50.4	65.4	58.9
School district	51.5	48.5	64.6	54.5	46.6	52.3	56.1	45.1	47.4	51.7
Energy-saving features	80.6	78.0	81.0	76.1	77.6	85.0	78.5	75.2	75.6	78.7
Recommendation of another buyer	12.6	12.1	11.6	15.9	17.2	16.8	15.9	15.0	10.3	14.5
Other	8.7	15.2	12.2	10.2	12.9	11.2	12.1	8.8	10.3	11.7
Number of respondents	103	132	147	88	232	107	107	113	78	1,107
Was your new home advertised as energy-efficient?										
Yes	34.5	44.2	40.6	48.9	50.8	41.7	58.5	57.0	28.7	45.8
No	57.1	45.6	40.6	34.8	38.0	47.5	33.9	32.0	66.7	42.8
Do not remember	8.4	10.2	18.8	16.3	11.2	10.8	7.6	10.9	4.6	11.3
Total	100.0	100.0	100.0	100.0	100.0	100.0	100.0	100.0	100.0	100.0
Number of respondents	119	147	165	92	250	120	118	128	87	1,226
What type of house would you prefer to buy?										
Single-family, detached	91.8	89.3	95.2	90.4	93.0	91.9	95.0	93.1	95.5	92.8
Single-family, attached	3.3	4.7	3.0	3.2	3.5	4.9	3.3	3.1	3.4	3.6
Duplex	0.0	2.7	0.6	4.3	0.8	1.6	0.8	0.8	0.0	1.2
Low-rise condominium	4.1	2.0	1.2	2.1	1.6	1.6	0.8	0.8	0.0	1.6
Mid-rise condominum	0.8	0.0	0.0	0.0	0.8	0.0	0.0	2.3	0.0	0.5
High-rise condominium	0.0	1.3	0.0	0.0	0.4	0.0	0.0	0.0	1.1	0.3
Mobile home	0.0	0.0	0.0	0.0	0.0	0.0	0.0	0.0	0.0	0.0
Total	100.0	100.0	100.0	100.0	100.0	100.0	100.0	100.0	100.0	100.0
Number of respondents	122	150	166	94	257	123	121	130	89	1,252

Source: NAHB, "Decisions for the '80s," pp. 16, 18, 53, 54, 59. Figures are rounded.

a. For a list of the states included in each census region, see table 1, footnotes a–i.

Table 3. *Willingness of Homeowners to Add Energy-Saving Features to Their Homes and the Types of Heating System under Consideration, by Census Region*[a]

Percent of respondents

Item	*New England*	*Middle Atlantic*	*East North Central*	*West North Central*	*South Atlantic*	*East South Central*	*West South Central*	*Mountain*	*Pacific*	*Total*
Willingness to pay $500 for extra insulation										
Quite willing	62.8	66.0	63.9	69.5	60.0	62.9	61.3	52.8	55.1	61.6
Somewhat willing	32.2	24.7	27.7	28.4	32.2	30.6	32.8	41.7	37.1	31.6
Not at all willing	5.0	9.3	8.4	2.1	7.8	6.5	5.9	5.5	7.9	6.8
Total	100.0	100.0	100.0	100.0	100.0	100.0	100.0	100.0	100.0	100.0
Number of respondents	121	150	166	95	255	124	119	127	89	1,246
Willingness to pay $1,200–1,500 for storm windows or windows with thermal panes										
Quite willing	56.2	52.0	42.8	45.3	36.6	46.0	46.7	44.4	46.1	45.2
Somewhat willing	34.7	31.8	49.4	47.4	36.6	41.9	36.7	42.1	33.7	39.3
Not at all willing	9.1	16.2	7.8	7.4	26.8	12.1	16.7	13.5	20.2	15.5
Total	100.0	100.0	100.0	100.0	100.0	100.0	100.0	100.0	100.0	100.0
Number of respondents	121	148	166	95	254	124	120	126	89	1,243
Homeowners considering installation of another type of heating system										
Yes	40.8	29.6	18.5	16.0	11.3	15.6	5.9	17.5	25.9	19.2
No	59.2	70.4	81.5	84.0	88.7	84.4	94.1	82.5	74.1	80.8
Total	100.0	100.0	100.0	100.0	100.0	100.0	100.0	100.0	100.0	100.0
Number of respondents	120	142	162	94	247	122	119	126	85	1,217

Homeowners converting or supplementing their heating systems										
Converting primary system	14.9	31.7	11.5	0.0	33.3	11.8	37.5	4.5	5.6	17.8
Adding supplemental system	85.1	68.3	88.5	100.0	66.7	88.2	62.5	95.5	94.4	82.2
Total	100.0	100.0	100.0	100.0	100.0	100.0	100.0	100.0	100.0	100.0
Number of respondents	47	41	26	13	27	17	8	22	18	219
Type of heating system under consideration										
Electric resistance system	5.8	2.0	0.0	0.0	2.5	0.0	11.1	0.0	0.0	2.2
Electric heat pump	3.8	24.5	11.4	11.1	30.0	10.0	44.4	0.0	11.5	15.1
Gas-fueled furnace	3.8	26.5	8.6	5.6	15.0	0.0	0.0	0.0	0.0	9.2
Oil-fueled furnace	0.0	4.1	0.0	0.0	0.0	5.0	0.0	0.0	0.0	1.1
Liquid petroleum system	0.0	0.0	0.0	0.0	0.0	0.0	0.0	0.0	0.0	0.0
Wood-burning system	63.5	28.6	60.0	66.7	22.5	75.0	0.0	78.3	69.2	51.5
Solar-powered system	7.7	10.2	14.3	11.1	17.5	5.0	44.4	17.4	11.5	12.9
Other	15.4	4.1	5.7	5.6	12.5	5.0	0.0	4.3	7.7	8.1
Total	100.0	100.0	100.0	100.0	100.0	100.0	100.0	100.0	100.0	100.0
Number of respondents	52	49	35	18	40	20	9	23	26	272

Source: NAHB, "Decisions for the '80s," pp. 64, 65, 67, 68, 69. Figures are rounded.
a. For a list of the states included in each census region, see table 1, footnotes a-i.

Table 4. *Characteristics of the Heating Systems of Housing Units Constructed between 1970 and 1978*
Percent of housing units

Characteristic	Period of construction 1975–78	Period of construction 1970–74	Period of construction 1970–78	Average, newly built single-family units, 1975–78
Fuel				
Natural gas	40	42	55	39
Electricity	45	35	16	50
Oil	8	13	22	8
Other	7	10	7	3
Type of system				
Warm-air furnace	64	67	50	80 (warm-air furnace and electric heat pump combined)
Electric heat pump	9	2	2	
Hot water or steam	1	3	16	5
Built-in electric	18	14	7	11
Other	8	14	25	5

Sources: U.S. Department of Energy, *Residential Energy Consumption Survey: Characteristics of the Housing Stock and Households*, DOE/EIA-0207/2, table 12B, p. 41; and U.S. Bureau of the Census, *Characteristics of New Housing, 1980*, Construction Report C25-80-13 (Government Printing Office, 1981), pp. 21, 23.

was of the energy characteristics of the existing stock of housing; the second was of the conservation activities actually carried out during the period April 1977 to December 1978.[1]

CHARACTERISTICS OF THE HOUSING STOCK AND HOUSEHOLDS. As one example of the data on characteristics of the existing housing stock, table 4 shows the data of the U.S. Department of Energy (DOE) for characteristics of heating fuels and heating systems, and compares these data with similar values from data of the Bureau of the Census on characteristics of newly built units from their C25 reports.

The specific numbers from the DOE survey shown in table 4 give heating characteristics according to the period during which the house was built—1975–78, 1970–74, and 1970–78. The C25 values are averaged for the period 1975–78. The magnitude of the percentages is reasonably close, particularly in view of the fact that the DOE survey includes multifamily units. Also, the trends implied by the DOE survey are

1. See U.S. Department of Energy, *Residential Energy Consumption Survey: Characteristics of the Housing Stock and Households*, DOE/EIA-0207/2 (Government Printing Office, 1980); and DOE, *Residential Energy Consumption Survey: Conservation*, DOE/EIA-0207/3 (GPO, 1980).

Table 5. *Presence of Conservation Features in U.S. Housing, by Region, Climate, Period of Construction, and Income of Occupants, 1970–78*
Percent of housing units

Description	*Attic insulation*	*Wall insulation*	*Some storm windows*	*Some storm doors*
All United States	69	50	61	61
Region				
Northeast	65	50	91	78
North Central	76	56	90	85
South	67	49	34	47
West	63	43	26	27
Climate				
More than 7,000 HDD[a]	81	65	94	74
5,500–7,000 HDD	77	57	84	82
4,000–5,499 HDD	65	49	79	74
Fewer than 4,000 HDD[a]	62	42	23	33
Period of construction				
1975 or later	86	84	66	42
1970–74	74	69	57	52
1950–69	79	54	56	63
1949 or earlier	56	37	65	65
Income of occupants				
Less than $5,000	41	28	48	48
$5,000–14,999	62	46	56	59
$15,000–24,999	77	57	67	66
More than $25,000	86	63	70	69

Source: Department of Energy, *Residential Energy Consumption Survey: Conservation*, DOE/EIA-0207/3 (GPO, 1980), table 1A, p. 5; table 5A, p. 21; table 7A, p. 29.
a. Heating degree days (see text for definition).

consistent with those implied by the C25 survey. Electric heat pumps are an item of particular interest; the prevalance of this item is shown in the C25 reports only for single-family homes built since 1978. Between 24 percent and 26 percent of all such units have electric heat pumps. The DOE data indicate a strong trend toward the use of electric heat pumps: 9 percent of the units built between 1975 and 1978 contain the pumps, while only 2 percent of those built earlier had them.

The DOE survey also showed the percentage of units that contained storm windows and doors and attic and wall insulation. The survey results are cross-tabulated by regional location, regional temperature levels, age of the unit, and income level of the occupants. These data are summarized in table 5.

Of all the units in the country, 69 percent have attic insulation, 50 percent have wall insulation, and 61 percent have storm doors and some storm windows. The North Central region has the most intensive insulation and the West has the least, in most categories. The Northeast and the South are roughly equal in attic and wall insulation, but the South has fewer storm windows and doors.

With regard to temperature levels, the need for insulation is measured by heating degree days (HDD). An HDD is the average number of degrees daily by which the temperature is below sixty-five degrees Fahrenheit; hours in which the temperature is above sixty-five degrees Fahrenheit count as zero. If the temperature were thirty-five degrees steadily, this would count as 30 HDD for the day. Thus, higher ratings of HDD indicate a greater need for insulation. The data show a consistent pattern, with more intensive insulation in the higher HDD areas, with the exception of storm doors in areas that have more than 7,000 HDD.

The pattern of the amount of insulation in relation to the year of construction of the house varies with the type of insulation. Newer units generally have greater amounts of attic and wall insulation. With respect to storm windows the pattern is more complicated, and older units systematically use more storm doors. It is possible that better insulation in basic doors and windows has reduced the need for storm windows and storm doors in newer units.

Finally, in relation to income, the pattern is systematic with respect to each type of unit: higher income levels imply greater concentrations of insulation. Of course, income is correlated with other characteristics of housing— particularly the year of construction—so the relation between income and insulation may be spurious. There is some indication, however, that the financial ability to purchase insulation is an important determinant of its presence.

ENERGY CONSERVATION ACTIVITIES. In the conservation section of the Residential Energy Consumption Survey, data on conservation activities undertaken between April 1977 and December 1978 were tabulated (see table 6). The two main categories of conservation activities surveyed are "added insulation" and "added equipment." Each of these was separated into two further groups—"inexpensive" and "expensive." Some examples of the activities in each category are:

Inexpensive insulation: Weather stripping, caulking, plastic covering

Expensive insulation: Roof or attic, basement or crawl space, outside walls

Table 6. *Installation of Insulation or Equipment for Conservation of Energy in U.S. Housing, by Region, Climate, Period of Construction, and Income of Occupants, April 1977–December 1978*
Percent of housing units

Description	*Inexpensive insulation*	*Expensive insulation*	*Inexpensive equipment*	*Expensive equipment*
All United States	42	12	14	11
Region				
Northeast	47	14	12	12
North Central	50	14	17	12
South	40	9	15	10
West	26	10	9	6
Climate				
More than 7,000 HDD[a]	43	16	12	12
5,500–7,000 HDD	48	15	16	11
5,000–5,499 HDD	47	13	14	13
Fewer than 4,000 HDD[a]	33	17	12	18
Period of construction				
1975 or later	33	18	21	13
1970–74	40	10	14	8
1950–70	42	12	14	10
1949 or earlier	44	11	13	11
Income of occupants				
Less than $5,000	30	5	6	5
$5,000–14,999	42	10	11	8
$15,000–24,999	49	14	16	12
More than $25,000	43	16	20	15

Source: Department of Energy, *Survey: Conservation*, table 9A, p. 38; table 15A, p. 61; table 21A, p. 85; table 27A, p. 109.
a. Heating degree days.

Inexpensive equipment: Storm doors, automatic thermostat, new water heater

Expensive equipment: Insulating or storm windows, electric heat pump, new furnace.

Cross-tabulations of these data are available by the four primary census regions—Northeast, North Central, South, and West—by temperature level, by age of unit, and by income of occupants.

Inexpensive insulation was added in about 42 percent of the units in the country, while the other three were added in between 11 percent and 14 percent of the units. The fact that the survey covered only about two years makes the interpretation of these values difficult. On the one hand, activities such as the addition of inexpensive insulation are not much more than normal maintenance, so high values are to be anticipated.

On the other hand, the addition of expensive insulation and equipment would typically be done only infrequently, so addition of these in 10 percent of the units during a two-year period might well be comparable to their addition in 50 percent of the units during a ten-year period.

The disaggregations by region and temperature exhibit a similar pattern. Within the regional groups, the Northeast and North Central regions undertook more conservation activity than did the South and West. Also, the North Central was generally the most active and the West the least active in each category.

The two groups having the most HDDs generally show more conservation activity than the two that have the fewest, although there are more exceptions here. In particular, the group having fewer than 4,000 HDD actually exhibited the largest amount of expensive conservation activity; this could well represent the need for air conditioning in these areas and the growing use of heat pumps to satisfy the need for both air conditioning and a modest amount of heating.

The tabulation by year of construction provides interesting, sometimes confusing, patterns. It would be expected that the newer units would need less conservation activity than older units, but this pattern holds in only one category—the addition of inexpensive insulation. The oldest units, built in 1949 or earlier, actually show a relatively low level of activity in the other three categories. This may reflect the technical difficulty of carrying out conservation efforts in such units, or it may represent economic decisions based on the lower value of such units or the lower incomes of those who occupy them.

Finally, with regard to the income level of the occupants, a clear pattern appears consistently across all four types of conservation activity. In each case, occupants in the two highest income categories undertake the greatest amount of conservation activity. This might reflect the fact that those in the higher-income groups can afford to make the necessary capital expenditures.

Energy Aspects of the Construction Industry

Concerning the prices of houses, both newly built and existing, construction costs, and the energy component of construction costs, the main question is whether energy represents a major part of the cost of residential construction and has therefore been an important factor contributing to the escalation of house prices and the "affordability

Table 7. *Costs of the Hard Components of Construction, 1974 and 1979*

Description	*1974 (dollars)*	*1979 (dollars)*	*Percent of total, 1979*	*Percent increase, 1974–79*
Basement and masonry				
Excavation	248	509	1.7	105.2
Masonry	1,158	1,952	6.6	68.6
Concrete	1,446	2,325	7.9	60.8
Total	2,852	4,786	16.2	67.8
Structural				
Lumber	3,806	6,409	21.8	68.3
Carpenters' labor	1,840	2,558	8.7	39.0
Other[a]	4,023	6,910	23.5	71.8
Total	9,669	15,877	54.0	64.2
Mechanical and electric				
Electric	824	1,432	4.9	73.9
Plumbing	1,559	2,267	7.7	45.4
Heating	871	1,498	5.1	72.0
Total	3,254	5,197	17.7	59.7
Finishing				
Total[b]	2,265	3,541	12.1	56.3
Total cost of hard components	18,040	29,401	100.0	63.0

Source: National Association of Home Builders, Economics Division, "Construction Cost Data Components" (Washington, D.C.: NAHB, 1980).

a. Includes drywall, insulation, roofing materials, millwork, gutters, and hardware.

b. Includes tile, floor covering, painting, appliances, and incidental materials and expenses.

problem" for new home buyers. The failings of construction cost data are well known, so here I can do no more than suggest what the significant trends are.

Table 7 shows the costs of the hard components of residential construction—that is, the costs of the materials and labor used to construct the unit. They do not include the costs of the site, overhead, financing, or profits. The data are shown in four categories, with some additional detail, for the period 1974–79. In 1979 the total hard-component cost of the average home was $29,400 and had risen 63 percent since 1974. More than half the cost represents structural items with the associated labor, lumber and carpenters' labor representing well over a quarter of the total. Each of the other categories—basement and masonry, mechanical and electric, and finishing—represents about a sixth of the total cost. During the period 1974–79 carpenters' labor and plumbing costs inflated the least among the components and excavation costs the most.

Table 8 combines the costs of the hard components with the costs of

Table 8. *Components of Average Prices of New Homes, 1975 and 1979*

Component of cost	*1975 (dollars)*	*1979 (dollars)*	*Percent of total, 1979*	*Percent increase, 1975–79*
Hard components	18,645	29,401	49.0	57.7
Finished lot	9,682	15,000	25.0	54.9
Other[a]	13,449	15,549	26.0	15.6
Total	41,776	59,950	100.0	43.5

Source: Unpublished estimates of the National Association of Home Builders, Economics Division.
a. Includes builder's overhead, profit, financing, and marketing costs.

Table 9. *Cost of the Energy Used in Construction of a Typical Single-Family House, 1979*[a]

Component	*Cost (dollars)*	*Percent of total*
Lumber and related items	1,785	32.4
Concrete	853	15.5
Flat glass	35	0.6
Siding	70	1.3
Insulation materials	82	1.5
Other[b]	831	15.1
Construction	1,848	33.6
Total	5,504	100.0

Source: Unpublished estimates of the National Association of Home Builders, Economics Division.
a. Does not include energy used to transport raw materials to factory site or finished products to construction site.
b. Includes miscellaneous items such as paint, plumbing, roofing, millwork, and hardware.

the site and other items, including the builder's overhead, profit, financing, and marketing costs, to arrive at the total price of a new home. It can be seen that the hard components represent just about half the full price of a house, the finished lot and other costs each representing about a fourth. According to these data, the other items inflated much less than the costs of the hard components and the site during the period 1975–79.

Table 9 contains estimates made by the National Association of Home Builders of the costs of the energy used in constructing a typical single-family home in 1979. These energy costs, of course, occur primarily in the production of the raw materials and their assembly into the housing unit. They do not include the cost of transporting the raw materials to the factory or the finished products to the construction sites. They are based on a price of eighty-six cents a gallon of petroleum or its equivalent. The total energy cost is about $5,500. This represents less than 20 percent of the total cost of the hard components and less than 10 percent of the

Table 10. *Trends in Prices and Costs, 1974–80*

Description	1974	1980	Percent change, 1974–80
Consumer price index[a]			
All items	147.7	246.8	67.1
Home purchase	142.7	254.7	78.5
Fuel oil and coal	214.6	556.0	159.0
Gas and electricity	145.8	301.8	107.0
Producer price index[b]			
All commodities	160.1	268.8	67.9
Energy	208.3	574.0	175.6
Construction costs[c]			
Department of Commerce composite	81.8	143.3	75.2
E. H. Boeckh (residential)	79.4	128.9	62.3
House prices (dollars)			
New houses	38,900	76,300	96.1
Older houses	35,800	72,800	103.3

Sources: Bureau of the Census, *Statistical Abstract of the United States, 1981* (GPO, 1981), pp. 460, 467, 754; Bureau of the Census, *Price Index of New One-Family Houses Sold*, Report C27 (GPO, various issues); and unpublished data from the National Association of Realtors.

a. Prices paid by urban wage earners. 1967 = 100.

b. 1967 = 100.

c. 1977 = 100.

average home price, using the 1979 data from tables 7 and 8. With respect to the distribution of energy costs, lumber materials and construction activity represent about a third of the total energy use each; concrete is the only other substantial item identified.[2]

Overall, the data in tables 7–9 suggest that housing construction is not a particularly energy-intensive activity. While about a fifth of the costs of the hard components can be attributed to energy—which is certainly not small—the hard-component costs in turn represent only about half the final price of the house. Thus rising energy prices would not be expected to cause the construction costs of housing to rise significantly faster than the average economic item.

In table 10 a number of aggregate price indexes are compared in order to confirm this point. The data for 1974 and 1980 and the cumulative percentage increase during the period are shown. The first rows show the aggregate consumer price index (CPI) and three of its components. The CPI for all items increased 67 percent during the period, while the

2. Energy prices have risen faster than the overall inflation rate since 1979, and therefore the energy share of construction costs would be higher now.

home-purchase component increased somewhat more and the two energy components significantly more. The producer price index (PPI) shows a similar pattern for all commodities and for the energy component.

The construction-cost indexes increased at rates similar to the CPI and PPI for all items and therefore significantly less than their energy components. The prices of new homes increased about 96 percent and the prices of older homes about 103 percent. These increases are somewhat greater than the increases in all items but distinctly less than the increases in the energy components. Overall the data are consistent with construction costs and home prices that do not show any special sensitivity to energy prices.

The Costs of Homeownership

There has been a great deal of interest and concern over the costs of homeownership and the ability of people, particularly first-time home buyers, to afford units. Recent data from the FHA surveys of the characteristics of newly built and existing units financed under the Section 203 program are an excellent source for analysis of energy costs as a component of housing costs because they cover all relevant costs of homeownership and they are consistent with the passage of time. They do have the drawback, however, that FHA homes tend to be priced lower than the average, a tendency that may create a bias in the data.

Table 11 shows the average monthly costs of homeownership, in five categories, for both newly built and existing units. A look first at the percentage distribution of the expense categories for new units in 1980 makes it clear that debt service, at about 72 percent, is the principal component. Heating and utility expenses, at 11 percent, are the second largest. The comparable data for existing units show a similar pattern. Heating and utility expenses for existing units, however, are higher, at 13.5 percent.

In the percentage increase between 1974 and 1980, debt service ranks second behind hazard insurance, heating and utility expenses coming in third. Again, the results for newly built and existing units are similar, although heating and utility costs did rise faster for existing units.

Thus, debt service, by virtue both of its current percentage of the total and of its growth in recent years, ranks as the principal item of concern by a wide margin. Debt-service expenses are, of course, the joint product of house prices and the costs of financing, both of which

Table 11. *Average Monthly Costs of Homeownership, 1974–80*

Component of cost	1974 *(dollars)*	1980 *(dollars)*	*Percent of total, 1980*	*Percent change, 1974–80*
Newly built units				
Debt service	196.73	489.27	72.4	148.7
Hazard insurance	9.10	23.06	3.4	153.4
Real estate taxes	40.56	58.32	8.6	43.8
Maintenance and repair	19.08	30.59	4.5	60.3
Heating and utilities	38.00	74.73	11.1	96.7
Total	303.47	675.97	100.0	122.7
Existing units				
Debt service	169.38	412.77	69.7	143.7
Hazard insurance	6.01	18.26	3.1	203.8
Real estate taxes	33.45	49.66	8.4	48.5
Maintenance and repair	16.62	31.36	5.3	88.7
Heating and utilities	36.98	79.80	13.5	115.8
Total	262.44	591.85	100.0	125.5

Sources: U.S. Federal Housing Administration, *FHA Homes: Data for States and Selected Areas on Characteristics of FHA Operations under Section 203* (GPO, 1974); and ibid. (GPO, 1980).

have risen dramatically since 1974. Heating and utility expenses, while higher and rising more quickly than such other components as real estate taxes and maintenance, are clearly not at the core of the affordability problem.

It is true that the homeowner has greater control over heating and utility costs than over other costs. Indeed, it has been shown earlier that homeowners do appear to weigh energy costs carefully in their home-purchase decisions and are undertaking significant efforts to conserve energy. So it seems reasonable to conclude that as a potentially controllable item of housing expense, heating costs are important, and higher costs are likely to affect home-buying decisions. But the general issue of affordability is nonetheless dominated by the level of house prices and the costs of financing the units.

The Economic Choice between New and Older Units

The choice between newly built and existing units is important for home buyers, and given the comparative energy efficiency of the two classes, it has important implications for the energy characteristics of new units and for the retrofitting of existing units. By first considering the factors that determine the *relative* prices of newly built and existing units it is possible to draw out these implications.

Some of the factors that create differences between the prices of newly built and existing houses are the prestige and superior design characteristics of newly built units, the larger profit margins for builders on more expensive newly built units, and the longer life expectancy of newly built units. Each of these causes the prices of newly built units to be higher than the prices of existing units, even of units that are similar in all other respects. In the matter of prestige, it appears that newly built units command higher prices in part simply because they are new. This may have a rational economic basis, in that sellers of existing units may have more information concerning the quality of the units and may try to sell units with hidden flaws. In any event, whether the reason is economical or psychological, it does appear that newly built units command a price premium, even when everything else is identical.

Of course, newly built and existing units are generally not identical, and price comparisons of newly built and existing units are thus not between otherwise identical units. Newly built units are constructed, presumably, with design characteristics that correspond closely to the current tastes of consumers. Since tastes change, the design characteristics of previously constructed units are likely to be out of date, and the units would therefore sell at a discount. Energy use and conservation characteristics are, of course, a case in point, as is a lesser need for time-consuming maintenance activities.

Another fact that raises the relative price of newly built units is that builders' profit margins appear to be higher on more expensive units and builders thus tend to build these units. Consumers with tastes for such units and the means to purchase them thus find the newly constructed units desirable; consumers whose tastes and resources suggest the purchase of lower-priced units naturally look to the market for existing units. This separation leads to a filter-down process, in which newly constructed units are purchased by the wealthier consumers; as the units depreciate and age, they are passed down for purchase to lower-income consumers as existing units. Finally, another reason new units may command higher prices is that, being younger, they will last longer.

Taken together, these considerations indicate that the prices of newly built homes are likely to be higher than the prices of existing homes. This has generally been the case, and the data in table 10 for 1974 are typical in this respect. But as the table 10 data clearly demonstrate, since 1974 the prices of existing homes have risen faster than the prices of newly constructed homes. This trend in relative prices may be explained by an extension of the filter-down theory, taking into account the larger increase in all house prices in recent years and the affordability problems thereby created. As house prices have risen in relation to income, a greater percentage of potential home purchasers are forced to demand the lower-priced existing units. This shift in demand from newly built units to existing units naturally tends to compress the differential between the prices of newly built and existing units.

Another reason the prices of existing units may be rising in comparison to those of newly constructed units is that new units are smaller on the average than older units. There may also be other elements of quality that are not captured by data on price and age alone.

Overall, then, there are two sets of factors that affect the movement of the relative prices of newly built and existing homes. What might be called substitution effects, such as the better design characteristics of new units, point to a widening difference between the prices of newly built and existing homes, as the tastes and preferences of consumers or the economic incentives to look for certain features shift. What might be called income effects—those that arise from the question of affordability—favor existing units and thus compress the relative prices as long as the prices of existing homes are lower than the prices of newly built homes.

Rising energy prices enter into this scheme as factors that create both substitution effects and income effects. The substitution effect occurs because rising energy prices change the design characteristics that buyers desire in their homes; that is, to the extent that newly constructed units embody the currently cost-effective construction techniques for the conservation of energy, the spread between the prices of new and existing units will widen, and new construction will be stimulated. Of course, to the extent that existing units could be retrofitted with energy-conserving features, and to the extent that the capital costs of such retrofitting were competitive with the costs of building similar features into newly built units, the substitution effect would disappear. The evidence so far, however, indicates that the retrofitting of existing units, particularly the oldest units, is being done slowly.

The income effect of rising energy prices occurs through the rising cost of homeownership. The data presented in the preceding section indicate that energy costs have been among the more important components of the rising costs of ownership, although in both magnitude and growth, debt service remains the dominant component. To the extent that energy prices contribute to the affordability problem, demand will

Table 12. *Selected Characteristics of New Single-Family Houses Completed between 1974 and 1980*[a]
Percent, computed from unrounded figures

Characteristic	*Percent distribution*						
	1980	*1979*	*1978*	*1977*	*1976*	*1975*	*1974*
Total completed	100	100	100	100	100	100	100
Central air conditioning							
Installed	63	60	58	54	49	46	48
Not installed	37	40	42	46	51	54	52
Bathrooms							
1 bathroom	18	16	15	17	20	24	22
1½ bathrooms	10	11	11	13	13	17	18
2 bathrooms	48	48	48	47	45	40	40
2½ bathrooms	25	26	25	23	22	20	21
Bedrooms							
2 bedrooms or fewer	17	14	12	11	12	14	13
3 bedrooms	63	64	64	66	65	65	64
4 bedrooms or more	20	23	24	23	23	21	23
Exterior wall material[b]							
Brick	28	27	30	31	31	32	35
Wood or wood products	42	42	40	38	38	36	32
Stucco	13	13	13	12	12	10	10
Aluminum siding	9	11	11	11	11	11	12
Other types[c]	8	7	7	8	7	11	11
Fireplaces							
No fireplace	43	38	36	39	42	48	51
1 fireplace or more	57	62	64	61	58	52	49
Heating fuel[b]							
Gas	41	39	37	38	39	40	41
Electricity	51	51	52	50	48	49	49
Oil	3	7	8	9	11	9	9
Other types	5	3	3	2	2	2	1
Heating system[b]							
Warm-air furnace[d]	57	56	57	78	76	72	73
Electric heat pump	24	26	25	n.a.	n.a.	n.a.	n.a.
Hot water or steam	4	5	5	5	6	6	6
Built-in electric	9	10	10	12	14	16	17
Other types	6	4	4	5	4	5	4
Parking facility							
Garage: one car	13	12	12	14	13	14	16
Garage: two cars or more	56	62	62	60	59	53	52
Carport	7	7	7	7	8	9	10
No garage or carport	24	20	18	19	20	24	22
Square feet of floor area							
Less than 1,000	7	5	5	5	7	8	7
1,000 to 1,199	14	13	13	14	15	17	17
1,200 to 1,599	29	28	28	30	29	30	29
1,600 to 1,999	22	23	24	38	37	34	34
2,000 to 2,399[e]	13	15	16	n.a.	n.a.	n.a.	n.a.
2,400 and more	15	15	14	13	12	11	13

Sources: Bureau of the Census, *Characteristics of New Housing*, 1980, and earlier issues.

n.a. Not available.

a. Houses for which characteristics were not reported have been proportionally distributed to those for which the data were reported.

b. Data reflect only principal type.

c. Includes asbestos shingles, cinder block, stone, and other types.

d. Before 1978, includes heat pump.

e. Before 1978, houses with 2,000 to 2,399 square feet were included with those having 1,600 to 1,999 square feet.

be shifted to lower-priced units, which tend to be existing units, thereby narrowing the difference in price between newly constructed and existing units.

In summary, rising energy prices affect the differences in price between newly built and existing units in several ways, and although the individual links can be identified, it is difficult to estimate the net effect of the offsetting factors quantitatively. The actual data on the price difference show the prices of existing homes rising in relation to those of newly built units. This could well be explained by a dominant income effect that shifts demand to lower-priced existing units. But the income effect, in turn, is primarily related to the cost of debt service, and energy prices are distinctly a factor of secondary importance.

Housing Construction Activity

Rising energy prices have also influenced the pattern of housing construction and the characteristics of the units built. The types of unit constructed, the energy-saving features incorporated, the density of housing development, and the regional pattern of housing starts have all been affected. By no means, however, have all the trends taken the directions that would be indicated if rising energy prices were the dominant concern, and it is often difficult to ascertain how great their influence actually is.

Characteristics of Newly Built Units

The Bureau of the Census and the Department of Housing and Urban Development (HUD) together have published comprehensive indicators of the characteristics of newly built housing and the way they have altered with the passage of time.[3] Those characteristics that are related to the use and conservation of energy are summarized in table 12.

At least three of the characteristics appear to run counter to energy-conservation principles. The percentage of new homes with centrally installed air conditioning, for example, has risen steadily, from 48 percent in 1974 to 63 percent in 1980. If the rising cost of electricity were the

3. Bureau of the Census, *Characteristics of New Housing*, Construction Reports C25, both a monthly and an annual publication.

Table 13. *Presence of Central Air Conditioning in New Houses, United States, 1974–80*

Location of house and status of central air conditioning	Percent distribution[a]						
	1980	1979	1978	1977	1976	1975	1974
All United States	100	100	100	100	100	100	100
Installed	63	60	58	54	49	46	48
Not installed	37	40	42	46	51	54	52
Inside SMSAs[b]	100	100	100	100	100	100	100
Installed	68	65	62	57	53	50	53
Not installed	32	35	38	43	47	50	47
Outside SMSAs[b]	100	100	100	100	100	100	100
Installed	52	51	50	48	42	39	40
Not installed	49	49	50	52	58	61	60
Northeast	100	100	100	100	100	100	100
Installed	29	26	20	17	13	13	16
Not installed	71	74	80	83	87	87	84
North Central	100	100	100	100	100	100	100
Installed	45	47	47	44	40	35	36
Not installed	55	53	53	56	60	65	64
South	100	100	100	100	100	100	100
Installed	84	85	84	80	78	71	72
Not installed	16	15	16	20	22	29	28
West	100	100	100	100	100	100	100
Installed	47	46	42	36	29	29	34
Not installed	53	54	58	64	71	71	66

Sources: Bureau of the Census, *Characteristics of New Housing, 1980*, p. 5, and earlier issues. Figures are rounded.
a. Percentages were computed from unrounded figures.
b. Standard metropolitan statistical areas.

dominant factor, this percentage would be expected to decline. It is possible, of course, that the national trends shown are distorted by the changing geographic distribution of new construction. Table 13, however, shows the air-conditioning data disaggregated by location inside or outside an SMSA and by census region, and the same general trend is apparent in each disaggregated component. The same point is valid for all the other characteristics shown in table 12. It has also been suggested that the growth in air conditioning may be a side result of the greater use of heat pumps to conserve energy. The installation of heat pumps can explain some, but by no means all, of the growth in air conditioning.

The trends in the number of bathrooms and in the number of square feet of floor area also run counter to energy-conservation principles. The percentage of homes having two or more baths, for example, rose steadily from 1974 to 1979 and has remained essentially unchanged since then, while the percentage of homes with 2,400 or more square feet has followed the same trend.

The trends with respect to two of the characteristics, in contrast, at least since 1978, are consistent with the conservation of energy. The percentage of homes having two bedrooms or fewer and the percentage having no garage or carport, hence requiring less energy to build and possibly to heat, have increased. The trend toward fewer bedrooms, however, could simply indicate that there are fewer persons per household.

The data on the percentage of homes with one fireplace or more is even more difficult to interpret, since whether fireplaces conserve energy or increase its use depends on the type of fireplace and the use that is made of it.

Finally two other sections of the data have to do with the type of heating fuel and the type of heating system used. There are a distinct trend away from heating oil, a significant trend to other types of fuel, and a modest trend toward gas and electricity. These data are the most suggestive in the set with respect to the price incentives that affect housing energy decisions. The price of heating oil has risen substantially more than the prices of other sources of energy for residential use, while the prices of other sources, which presumably include both wood and solar energy, have doubtless risen the least. Among heating systems, there has been a distinct trend away from hot water and steam and from built-in electric heaters and a trend toward warm-air and other types of furnace. Somewhat surprisingly, there has been a slight downturn in the installation of electric heat pumps since 1978, when the data began to be tabulated.

The percentage of units using natural gas is of special interest in view of the potential for large price increases if it is deregulated in the future. The percentage of units using natural gas declined between 1974 and 1978, probably reflecting the limited supply of new connections available during this period. Between 1978 and 1980, as new connections were again available, the percentage rose, presumably because of the continuing price advantage.

Overall, the trends in the characteristics of new homes give a diverse picture of the effects of high energy prices. The distinct decline in the use of heating oil and natural gas is the one trend in which the results are clearly consistent with price incentives, although even here it appears that availability, as much as price, is the primary concern. With respect to the other characteristics, there are no distinct common trends; some characteristics move in accord with energy conservation, others in the opposite direction. Also, in essentially all the cases, factors such as household size seem to be important, in some cases dominant.

The distinction between the energy use of a housing unit and its energy efficiency may be a significant reason that the trends in characteristics are sometimes confusing. *Efficiency* is defined here by the energy required per unit of housing services. *Energy use* is then the product of efficiency and the level of housing services. Energy use can thus be reduced either by improving efficiency or by reducing the level of housing services consumed. Consumers seem to have been reducing their use of energy by increasing efficiency, but not by reducing the level of housing services consumed. Indeed, if the gains in efficiency are great enough, the consumer may then be able to afford higher levels of housing services.

The propensity for new homes to be purchased by higher-income households—generally not first-time buyers—is also important for the distinction between efficiency and use. Higher income and the capital gain realized from the sale of a previous home make it unlikely that these buyers will choose to reduce the level or quality of housing services consumed as a way of adjusting to higher energy prices, but they may well be attracted to greater energy efficiency, even if it requires an initial investment that brings returns only through an extended period.

Housing Starts, by Type of Structure

The structural aspects of housing units can significantly affect the energy used for space heating, construction costs generally, and land use. On all scores, multifamily units are the most efficient, single-family units the least efficient. Attached single-family units, moreover, are clearly more efficient than detached single-family units. The energy efficiency of various groups of multifamily units is more difficult to order. Generally, however, structures having five or more units are more efficient on all scores than those having two to four units.[4]

Table 14 shows U.S. housing starts, by type of structure, as percentages of the corresponding annual totals since 1974. In the first row of the

4. It has been suggested that a trend toward larger multifamily units would reduce the energy advantage of these units. The data on the size of multifamily units, however, do not reflect a clear trend, and the energy use of multifamily units would still remain distinctly lower than that of single-family units.

Table 14. *U.S. Housing Starts, by Type of Structure, 1974–80*[a]
Percent of corresponding total

Item	*1974*	*1975*	*1976*	*1977*	*1978*	*1979*	*1980*
Single-family homes as percent of total	66.4	76.9	75.6	73.0	70.9	68.4	65.9
Attached single-family homes as percent of all single-family homes	7.4	4.6	4.7	5.1	6.0	8.2	9.2
Structures of two to four units as percent of total	5.1	5.5	5.6	6.1	6.2	7.0	8.5
Structures of five or more units as percent of total	28.5	17.6	18.8	20.8	22.9	24.6	25.6
High-rise structures as percent of structures of five or more units	21.9	12.2	9.0	10.4	13.9	17.6	21.1

Sources: Bureau of the Census, *Housing Starts*, various issues, Construction Reports, C20 series (GPO).
a. The data do not include mobile homes.

table, it can be seen that single-family starts as a percentage of total starts has declined steadily since 1975. The percentage of single-family units that are attached, shown in the next row of the table, has also risen steadily since 1975. Both these trends are consistent with the rising costs of energy, of construction generally, and of land. The multifamily-structure mix of housing starts also seems fully consistent with distinct trends since the mid 1970s toward energy-efficient, construction-cost-efficient, and land-efficient units.

Although the trends since the mid 1970s in the mix of structure types are consistent with energy price effects and certainly have made new units more energy conserving, it is not clear that high energy prices have necessarily caused the shift in the mix of structure types. During the 1960s and early 1970s, for example, a period of declining relative costs of energy, single-family units as a percentage of all housing starts declined.[5]

Regional Considerations

Higher energy prices, taken by themselves, affect the geographic distribution of new housing construction in several ways. From the standpoint of space heating and air conditioning, housing activity would

5. The factors that have been responsible for the various trends might be sorted out through multivariate analysis, but I have not obtained significant statistical results with the limited number of observations available.

Table 15. *Mean Heating and Cooling Degree Days, by Region*

Region	Mean heating degree days	Mean cooling degree days
Northeast		
New England	6,484	445
Middle Atlantic	5,779	730
North Central		
East North Central	6,258	806
West North Central	6,576	1,045
South		
South Atlantic	3,075	1,813
East South Central	3,379	1,692
West South Central	2,254	2,502
West		
Mountain	5,626	1,018
Pacific	3,308	574
All United States	4,757	1,135

Sources: U.S. Department of Commerce, National Oceanic and Atmospheric Agency, Environmental Data Information Service, *State Regional and National, Monthly and Seasonal Heating Degree Days* (Asheville, N.C.: National Climatic Center, 1978); and Department of Commerce, NOAA, EDIS, *State Regional and National, Monthly and Seasonal Cooling Degree Days* (National Climatic Center, 1978).

be expected to shift toward regions where the energy required to moderate extreme temperatures would be least. Similarly, any changes in the relative prices of energy, whether oil, gas, or electricity, by way of its effect on both space heating and other household uses of energy, would influence regional housing choice. Finally, shifts in the regional location of manufacturing jobs, whether also caused by the rising cost of energy or by other factors, influence the location choices of households.

While in principle these three factors could point in different directions, in fact they tend to reinforce one another. Specifically, they are all consistent with strong movements from the Northeast to the South and West and with a more modest but still distinct movement from the North Central region to the South and West. These migration trends during the 1970s, frequently referred to as movements from the Snow Belt to the Sun Belt, have of course a variety of causes other than energy prices, but it is noteworthy that rising energy prices are at least consistent with these movements.

REGIONAL MEASURES OF THE EFFECTS OF ENERGY PRICES. There are several data sets available that demonstrate the importance of energy prices in the regional choice of housing. The first set, shown in table 15, is the number of heating and cooling degree days in the regions of the

United States. Several aspects of the weather data in table 15 stand out. First, the geographic variation in temperatures is quite dramatic, the region with the most heating degree days having nearly three times as many as the region with the fewest and the region with the highest number of cooling degree days having more than five times as many as the region with the fewest. Indeed, no single region has close to the United States average of 4,757 heating degree days, and only two regions have within 30 percent of the United States average of 1,135 cooling degree days. Second, in degree days, days that require heating vastly exceed cooling days. The lowest number of heating degree days, for example—2,254 in the West South Central region—is only slightly smaller than the greatest number of cooling degree days—2,502, also in the West South Central region. Third, regions that have a large number of degree days on one measure tend to have a small number on the other measure; the correlation is $-.74$. There are some exceptions, however: on both measures, for example, the West North Central region has a large number of degree days, the Pacific region a small number.

Two implications regarding the demand for energy can be drawn from the data on underlying weather conditions. First, the regions that have a large number of heating degree days, not offset by a lower than average number of cooling degree days, the Northeast and North Central regions and possibly the mountain area in the Western region, are clearly at a disadvantage. Second, the Pacific region seems to have the conditions closest to ideal, with only a few more heating degree days than the region with the fewest and only slightly more cooling degree days than New England. The Southern region follows, with no clear choice among its three subareas.

In order to determine the probable effect of the number of degree days on choice of regional location, it is necessary to compare heating technology and cooling technology, regional variations in energy prices, and the opportunities for conservation that are available to offset the amount of energy needed. It is difficult to make a simple comparison between the cost of heating a unit one HDD and cooling the same unit one CDD because the two problems frequently call for different technologies and different fuel mixes and because other, related environmental features may be relevant. Specifically, while it could be said that the energy needed to heat a unit one HDD and to cool the same unit one CDD with the same electric heat pump is about the same, factors such as the amount of sunlight, which would reduce the heating cost, and the humidity, which would increase the cooling cost, may be significant. The comparison becomes even more difficult when it is recognized that the furnaces used for most space heating are less efficient than most air conditioners.

Questions concerning regional variations in energy prices and conservation measures that can be taken can be considered more concretely. The Residential Energy Consumption Survey of the DOE is a convenient source of data. Table 16 shows data on energy consumption, expenditure, and implied prices for the average U.S. household and then shows the relative values disaggregated by region, year of construction, and income. During the period April 1978 to March 1979, the average U.S. household in a single-family detached unit consumed 158 million British thermal units (Btu) of energy, excluding wood and coal and energy for transportation, at an annual cost of $825, implying a price of $5.22 per million Btu.[6]

Regionally the patterns of consumption follow the expected lines, the Northeast and North Central regions about 25 percent higher than the national average and the South and West similarly lower than average. Since the implied price of energy in the Northeast is close to the national average, its expenditures are at the same extreme indicated by its consumption. The implied price per unit of energy in the North Central region, however, is relatively low, and that in the South is high, so their expenditures are closer to the national average than consumption alone would indicate. Both consumption and expenditure are low in the West. The variation in implied price by region is a function of both regional price variations for each type of fuel and regional variations in the mix of fuels used. This feature of the data could well be explored further.

As would be expected, the oldest units, built in 1939 or earlier, use the most energy. Beyond these oldest homes, the relation between consumption and age of the unit is generally as expected, but with several inversions. The reason may be that age of the unit is likely to be correlated with both its geographic location and the income of the household. It is also noteworthy that the range of expenditures by age of the unit is quite limited, since the implied price tends to vary inversely with the level of consumption.

6. A conversion factor is needed to translate residential electricity use into its Btu equivalent. The Residential Energy Consumption Survey uses a reasonable procedure for this conversion, but it should be recognized that there are alternative factors that could be used. It is unlikely that alternative calculations would change any of the conclusions drawn in the text.

Table 16. *Consumption of and Expenditures for Energy, Detached Single-Family Homes, by Region, Location, Year of Construction, and 1977 Income of Occupants, April 1978–March 1979*

Description	*Consumption in relation to that by the average U.S. household*	*Expenditure in relation to that by the average U.S. household*	*Implied price in relation to that paid by the average U.S. household*
Region			
Northeast	1.25	1.30	1.04
North Central	1.22	1.08	0.89
South	0.73	0.92	1.26
West	0.84	0.66	0.79
Location			
Inside SMSAs[a]	1.06	1.04	0.98
Outside SMSAs[a]	0.89	0.93	1.04
Year of construction			
1939 or earlier	1.13	1.03	0.91
1940–49	0.94	0.90	0.96
1950–59	1.01	1.02	1.01
1960–69	0.94	1.01	1.07
1970–74	0.84	0.98	1.17
1975 or later	0.86	0.98	1.14
1977 income of occupants			
Less than $5,000	0.75	0.68	0.91
$5,000–9,999	0.87	0.82	0.94
$10,000–14,999	0.87	0.91	1.05
$15,000–19,999	0.97	1.01	1.04
$20,000–24,999	1.06	1.07	1.01
$25,000 or more	1.22	1.23	1.01

Addendum	*Consumption (millions of Btu)*	*Expenditure[b] (dollars)*	*Implied price (dollars per million Btu)*
Average U.S. household	158	825	5.22

Source: Department of Energy, *Residential Energy Consumption Survey: Consumption and Expenditures, April 1978 through March 1979*, DOE/EIA-0207/5 (GPO, 1980), table 10, pp. 63, 65, 66.

a. Standard metropolitan statistical areas.

b. Excludes household use of gasoline for transportation and use of wood or coal.

As expected, consumption of energy does rise monotonically with income level. The range of variation is, in fact, actually greater than that caused by geographic location. Since the implied prices also tend to rise with income, the average expenditure per household rises dramatically with income.

In order to determine the effectiveness of energy-conservation activities, it is useful to consider the energy-consumption data by regions

Table 17. *Residential Energy Consumption, by Number of Heating Degree Days and Cooling Degree Days*
Millions of Btu

Number of heating degree days and cooling degree days	*Average household consumption of energy*
Fewer than 2,000 CDD and more than 7,000 HDD	162
5,500–7,000 HDD	167
4,000–5,499 HDD	158
Fewer than 4,000 HDD	99
More than 2,000 CDD and fewer than 4,000 HDD	91

Source: Department of Energy, *Survey: Consumption and Expenditures*, table 1, p. 12.

differentiated by heating and cooling degree days. Were no additional conservation of energy carried out in the extreme climates, energy use would be expected to vary proportionately with heating and cooling degree days. But energy conservation can be achieved either through capital investment in appropriate equipment and materials or through acceptance of less comfortable conditions.

The relevant data are shown in table 17. The data indicate that conservation of energy can significantly offset the demand for energy that would arise because of climatic conditions alone. In areas with more than 7,000 HDD, for example, less energy per household is actually used than in areas with 5,500 to 7,000 HDD. Similarly, in areas with 5,500 to 7,000 HDD only about 6 percent more energy is used than in areas with 4,000 to 5,500 HDD, although the HDD ratio based on the midpoint of the range would suggest consumption 32 percent greater. It is more difficult to make such calculations for the bottom categories because the midpoint of the range would not be an accurate measure.

Overall, the effectiveness of energy conservation seems evident. Even so, the range of energy consumption between the extremes of climatic conditions is still dramatic, and these calculations, of course, do not include the capital costs of the necessary equipment and materials to carry out this conservation. Finally, it must be recognized that such energy conservation is achieved in important part by accepting distinctly lower levels of comfort, and this must be, in the end, a significant cost of high energy prices.

THE REGIONAL DISTRIBUTION OF HOUSING STARTS. Data on the

Table 18. *Distribution of the U.S. Population among the Four Primary Census Regions, according to the 1980 Census*

Region	*Percent*
Northeast	21.7
North Central	26.0
South	33.3
West	19.0

Source: Bureau of the Census, *Statistical Abstract of the United States, 1981*, p. 12.

number of housing starts between 1974 and 1980, by four large census regions, by total housing starts, and by four structure categories are given in table 19. As a benchmark, the percentage of the national population contained in each of the four census regions from the 1980 census data is given in table 18. Comparing the distribution of the population with the distribution of total starts in 1980 from table 19, it is immediately clear that the Northeast and North Central regions are producing a smaller proportion of housing starts than would be indicated by population alone, while the Southern and Western regions produce more by the same standard. The trends in total housing starts between 1974 and 1980, moreover, are in the same rank order, with the Northeastern share of housing starts falling about 30 percent and the Southern share rising about 20 percent. These patterns are at least consistent with the regional climatic conditions and energy prices as reviewed in the foregoing section.

Table 19 also shows the regional distribution of housing starts by type of structure. Earlier I discussed the national trends in structural types. At the national level, the data indicated a strong trend away from single-family structures—although with some substitution to attached single-family units—and within the rising multifamily sector, strength in two- to four-unit structures, in structures having five or more units, and particularly in high-rise structures. The regional trends shown in table 19 never run contrary to the national trends. High-rise units, for example, have been relatively strong in both the Northeast and North Central regions, consistent with the high energy efficiency of these units. Single-family detached units have been relatively strong in the West, while attached single-family units and two- to four-unit structures have been relatively strong in the North Central and Southern regions.

The regional distribution of housing construction activity can also be considered according to its distribution within or outside standard metropolitan statistical areas (SMSAs). The available data, shown in table

Table 19. *Regional Housing Starts as Percentages of Total U.S. Housing Starts, by Type of Structure, 1974–80*

Region and type of structure	1974	1975	1976	1977	1978	1979	1980	*Percent change, 1974–80*
Northeast								
Total starts	13.7	12.9	11.0	10.1	9.9	10.2	9.7	−29.2
Single-family, detached	13.5	12.5	10.8	10.5	10.1	9.8	9.9	−26.7
Single-family, attached	12.1	14.6	14.5	16.2	12.8	17.3	12.8	5.8
Two to four units	13.5	12.8	9.6	7.9	8.2	7.3	7.7	−43.0
Five or more units, low-rise	10.9	12.3	9.3	7.4	7.9	5.6	5.1	−53.2
Five or more units, high-rise	26.9	27.7	36.4	19.4	18.0	34.9	24.0	−10.8
North Central								
Total starts	23.7	25.3	26.0	23.4	22.3	20.0	16.9	−28.7
Single-family, detached	26.1	25.3	25.6	23.7	23.1	21.2	17.1	−34.5
Single-family, attached	16.7	12.1	20.0	14.9	16.3	11.2	12.8	−23.3
Two to four units	22.5	26.6	26.9	23.5	22.6	21.5	18.4	−18.2
Five or more units, low-rise	23.2	28.2	28.2	23.5	20.5	19.6	18.4	−20.7
Five or more units, high-rise	9.5	17.7	33.7	26.9	25.4	14.1	11.5	21.1
South								
Total starts	41.3	38.1	37.0	39.4	40.8	42.8	49.7	20.3
Single-family, detached	41.7	41.6	40.4	40.8	42.4	44.2	50.1	20.1
Single-family, attached	37.9	29.3	30.9	35.1	38.4	38.8	51.3	35.4
Two to four units	32.1	25.2	21.5	26.4	27.8	34.0	45.4	41.4
Five or more units, low-rise	42.0	29.6	31.1	39.5	40.4	44.4	49.7	18.3
Five or more units, high-rise	46.4	27.7	15.7	37.7	37.8	35.8	50.7	9.3
West								
Total starts	21.3	23.7	26.0	27.1	27.0	27.0	23.7	11.3
Single-family, detached	18.8	20.7	23.3	25.1	24.5	25.0	22.7	20.7
Single-family, attached	33.3	39.0	34.6	33.8	32.6	31.6	24.4	−26.7
Two to four units	34.6	36.2	42.6	41.8	40.2	37.3	27.4	−20.8
Five or more units, low-rise	23.4	29.7	31.3	29.7	31.4	30.4	27.0	15.4
Five or more units, high-rise	18.2	26.9	14.2	16.0	18.7	15.3	13.6	−25.3

Sources: Bureau of the Census, *Housing Starts*, various issues.

Table 20. *Percentage of U.S. Housing Starts That Were in Standard Metropolitan Statistical Areas, 1974–80*[a]

Year	Percent
1974	69.0
1975	65.5
1976	67.9
1977	69.3
1978	70.9
1979	71.1
1980	70.7

Sources: Bureau of the Census, *Housing Starts*, various issues.
a. Privately owned units only.

20 as the proportion of housing starts that were in designated SMSAs, cover only the United States as a whole. It can be seen that the proportion of housing starts in SMSAs expanded significantly between 1975 and 1979 and remained close to the peak level in 1980. Although these results appear to be consistent with energy-conservation principles, more detailed data that would provide a cross-tabulation with regions of the country and types of structure are not available.

Summary

The evidence surveyed in this paper indicates clearly that consumers are keenly aware of the implications of rising energy prices for residential energy use. Consumers anticipate that the energy situation may become even worse and indicate a strong preference for energy efficiency in their housing units. Efforts to retrofit existing units so as to conserve energy are reasonably well documented. Two usual ways are the addition of insulation and supplementary heating systems. The principal supplementary heating systems added are wood-burning stoves, electric heat pumps, and solar devices.

The particular energy-conserving techniques used in newly constructed units are more difficult to document, and the technology itself was not a concern of the paper. Items such as more and better insulation, greater attention to careful construction and placement of fittings, location of the site, and direction of the unit appear to be factors that bring surprisingly high returns. More exotic technologies, such as solar-energy collection, are of minor importance.

For the trends toward energy conservation among the traditionally measured characteristics of newly constructed housing units, the results are mixed. Progress seems to have been made with some measures, such as reductions in the number of bedrooms. With other measures, the trends, such as to units with centrally installed air conditioning, go against energy conservation. Overall consumers seem not yet to have been willing to modify many of the basic characteristics of housing units, such as size and the number of bathrooms, in the effort to conserve energy. The greatest progress has been in additional expenditures for insulation and equipment that reduce energy use but that have no effect on the services derived from the unit.

This consumer behavior emphasizes the distinction between the energy use of a housing unit and its energy efficiency. Households appear to be unwilling to reduce their use of energy by consuming fewer housing services, but they do seem willing to increase energy efficiency when the financial considerations are favorable and the use of the housing is not impaired.

Strong regional differences appear in the energy efficiency of existing units, in current efforts to improve their efficiency further, and in regional patterns of housing construction activity. Given the strong regional variations in heating and cooling degree days, the relatively modest regional variations in energy use are testimony either to the use of energy-efficient techniques in the regions where climates are harshest or to willingness to accept less comfortable homes in these regions. It is difficult to differentiate these two forms of energy conservation by means of the available data. But despite these ways of offsetting the underlying climatic conditions, the shift of housing construction activity from the Northeast and North Central regions to the South and West is quite impressive. Unfortunately, it is not easy to separate the influence of residential energy in this Snow Belt to Sun Belt migration from that of a variety of other economic and social factors. The pattern of expenditures for energy, however, is clearly consistent with these findings concerning migration.

Type of structure provides another approach to energy conservation. Single-family detached units tend to be the least energy efficient; attached single-family units, two- to four-unit structures, and those having five or more units, in that order, are more efficient. The trends in volume of housing construction by type of structure since 1974 are at least consistent with efforts at this form of energy conservation.

Energy prices also influence the trade-off between new and older housing units. As the preferred construction techniques and desired design features change on account of energy conservation, there should be a substitution in demand to newly constructed units and away from existing units. Simply put, existing units become economically obsolete because of changes in the price of energy and related shifts in demand. This substitution effect tends to raise the prices of new homes in relation to the prices of older homes and to spur new construction. On the other hand, rising energy prices add to the affordability problem in the purchase of a home, although the rising cost of debt service is still by far the largest component of the rising cost of homeownership. Affordability problems tend to shift demand from newly built units to existing units to the extent that existing units are lower in price. In recent years, such income effects have seemed dominant as the prices of existing homes have risen significantly in relation to those of newly constructed homes. This result is also consistent with the fact that rising energy prices do not appear to affect the construction cost of new units with any special severity.

It is difficult to translate the data on recent developments that have been reviewed in this paper into a forecast of future trends. For one thing, the outlook for the trend in real energy prices is itself substantially more complicated today than it seemed a year or two ago. For another, it is clear that the response of both the demand for housing and the supply of it to high energy prices is a dynamic adjustment, and it is difficult to know how much further change is already on the way. Third, factors other than energy impinge significantly on almost all aspects of housing construction. General problems of affordability caused by rising interest rates, for example, have certainly affected the demand for housing more significantly in recent years than have rising energy prices. On the other hand, it is quite possible that were interest rates to moderate, conservation of energy might receive more attention.

Despite three good reasons for not attempting to forecast future developments in housing construction that will be influenced by high energy prices, I am still willing to give my impressions briefly of the probable direction of these developments. It should be stressed, however, that these are intuitive impressions rather than systematic results of the material reviewed in the paper.

Further progress toward energy efficiency in U.S. housing should be anticipated in improved insulation and equipment simply on the basis of their current cost effectiveness. If energy prices were to remain at high levels, and certainly if they were to rise further, more innovative techniques and products would be likely to come into use. Energy prices are the key variable in determining whether these more advanced technologies will come into play; for the most part the devices have already been developed.

Three categories can be used to distinguish the various technologies available for increasing residential energy efficiency.[7] Current practice includes the improvements in equipment and insulation already evident in new construction. The second category is highly energy efficient; it includes techniques with well-documented results that require only price incentives to ensure widespread adoption. Techniques in the third category, extreme designs, are unlikely to be adopted, because of considerations of cost and reliability, in the absence of major price shocks or technological advances.

The highly energy-efficient modes that can be anticipated for adoption in the near future include air-exchange systems, microelectronic controls, and a variety of passive solar-energy systems. Such techniques can reduce the use of energy by 75 percent or more. Extreme designs include superinsulation or double insulation, active solar-energy systems, and individual wind systems. While such systems are capable of virtually eliminating the use of nonsolar energy, general application of them is still far in the future.

Although adoption of new technology in the highly energy-efficient category to a significant extent is likely, most households will not trade the basic services derived from housing for improvements in energy efficiency. I do not foresee, for example, significant reductions in the average size of American homes because of high energy prices. One reason is that newly produced units are typically purchased by higher-income households, who have less of an economic incentive to reduce the level of services that they consume. Lower-income households, on the other hand, who might well be prepared to make such a trade-off, generally find older units more affordable; but the costs of major structural changes and retrofitting for energy efficiency in older units seem too high, at least from the current vantage point. This, of course, is not to

7. See William Marcuse, "Energy Pricing and Availability, 1980–1990," David J. MacFadyen, "Housing, Transportation, and Energy Systems Technologies that Relate to Housing Markets," and Larry E. Ruff, "Housing and Energy in the 1980s," papers presented at a Brookings Institution conference, Housing and Energy in the 1980s, October 23, 1980.

deny that major changes in the characteristics of U.S. housing may come about in such forms as distinctly smaller units. But current energy prices appear as only one of a number of factors that would provide incentive for such changes.

The structure-type mix of U.S. housing will also be influenced by energy prices, but energy prices at current price levels do not appear to be the dominant factor. Specifically, I anticipate a continuation of the trend away from the detached single-family unit, but I suspect that it will come about more because of the high cost of housing and mortgage financing and the difficulty of builders in finding the large expanses of land necessary for development of traditional single-family detached units.

The regional distribution of housing construction could well be the source of the most surprises during the coming decade. The current trends toward construction in the South and West are clearly consistent with energy-conservation principles. But they are also consistent with many other social and economic forces, and it is difficult to evaluate the importance of energy prices as a causative factor in this process. To forecast the future regional distribution of construction is even trickier in view of the strong changes in the relative prices of housing across geographic regions. At some point, for example, it seems likely that the large increases in California housing prices in relation to prices in the Northeast and North Central regions will significantly curtail the flow of migration to California. From an economist's standpoint, some stabilization and even reversal of these migration patterns seems likely to be the result of the changing relative prices, but when this will happen is less clear; it could happen in the 1990s rather than the 1980s.

Finally, I do not foresee high or even moderately rising energy prices significantly affecting the volume of new housing starts in the next decade. The data presented here on the costs of homeownership indicate that only 11 to 13 percent of the present costs of homeownership can be attributed to heating and utility costs. Were the real price of energy to rise even at rates of 2–3 percent annually, the share of energy in the costs of homeownership would rise only to 14–17 percent by the end of the decade.

Deregulation of natural gas prices is an event that could lead to greater increases in the cost of residential energy. The effect of an increase in the price of one source of energy, however, will be reduced by the ability of consumers to substitute other sources. Also the availability of natural gas would be increased at higher prices, and since it appears that new connections for gas are, or have been, limited by lack of availability, greater availability would offset some of the price effect.

While various increases in costs are thus not trivial, they tend to seem mild in a world in which rising house prices and mortgage interest rates can easily raise homeownership costs 5 or 10 percent in a single year. I therefore see the conflict between high levels of housing demand attributable to demographic factors and the affordability dilemma for first-time home buyers as the primary force determining levels of housing production during the 1980s, with energy prices influencing them much less.

Comment by Robert J. Sheehan

In general, I support the conclusions of Professor Jaffee. But he did not cover some additional factors that support his position or raise some significant issues that may affect the decisions of consumers in the future.

First, he referred to the NAHB survey of new home purchasers for units purchased in 1977 and 1978. The survey was undertaken late in 1980. The responses were therefore obtained during the period when gasoline prices were rising rapidly as a result of the problems in the Middle East. The respondents were asked to describe the preferred characteristics of their next home purchase. This is the trade-up market of the 1980s. Most said they wanted larger houses and were willing to travel an extra half hour to have them, so they are not minimizers of energy use. Their housing expectations really had not changed, even during a period when energy prices were rising rapidly. They were also asked extensive questions concerning energy and what trade-offs they were prepared to make.

The basic limitation of this survey is that first-time home buyers were ignored. In the future, their attitudes will have to be examined, especially in view of the general economic environment, the deregulation of financial markets, and the higher real capital costs of housing in the 1980s. Perhaps energy costs that influence the marginal purchaser's response could make a difference.

The overall response of the market to such costs is really unknown. The National Association of Home Builders (NAHB) is now attempting to formulate a questionnaire that will deal with renters, who are potential first-time buyers, but it is having difficulty obtaining a good mailing list. Lists of homeowners are easy to obtain, because the NAHB has a subsidiary corporation that insures new homes. But it is also desirable to survey people who may prefer to rent and renters in general.

Ultimately, no one really knows how the responses of consumers to energy costs will affect the consumption of housing in the 1980s. It is possible that consumers' expectations about housing may be about to change. No one knows whether home buyers will be able to realize their current expectations or how those expectations may change. This is especially true concerning the location of employment. Job location is critical to home construction in the 1980s because builders tend to follow employment. They do not create primary jobs except in those few isolated situations in which they are building retirement housing or recreation areas.

Perhaps the NAHB will make some surveys that will make it possible to understand the housing expectations of the whole market in the 1980s better. If there is an industry left to support it, the NAHB will conduct such surveys.

Another thing I want to emphasize is that better evaluations of what happened in the 1970s are needed. A large number of the existing units were heated by natural gas, which had not yet been deregulated. Deregulation of natural gas will cause a real increase in the costs of housing for a large part of the housing stock. It will therefore have a marked effect on housing markets in the 1980s. The effect upon heating costs may be awesome, since natural gas is used much more for heating than for air conditioning. Higher heating costs in cold climates could reinforce the migration patterns that appeared in the 1970s. That might intensify the general migration to the Southwest and the West, especially the Southwest. It already appears to be one of the fastest growing regions of the country.

Deregulation of natural gas will also affect trade-offs between new and older housing units. It could be a depressant to the prices of older homes. But it will also make units without natural gas much more attractive because of the high capital costs of conversion in both new and older homes. During the 1970s, builders in a growing number of areas relied mainly on electricity because of moratoriums on natural gas connections. So in those areas the newer housing stocks will have some competitive advantage as natural gas is deregulated.

For these reasons, it may be misleading to rely primarily on the way people responded to increases in energy prices in the 1970s in predicting what they will do in the 1980s. Not only operating costs but also the capital costs of insulation must be examined. Natural gas is used in the manufacture of a large share of the insulation in newly constructed or existing homes. Thus, I see deregulation of natural gas as a factor that must be taken into account and its effects carefully measured.

Jaffee ignored one energy characteristic of newly constructed homes—the fact that two-story homes tend to be somewhat more energy-efficient than one-story homes. During the last half of the 1970s, there was in all regions a definite increase in the number of two-story detached homes constructed. This is a small factor, but it could become significant in the future, especially given the trend toward smaller lots.

One problem with Jaffee's paper is its failure to deal in any detail with the characteristics of new multifamily homes. Jaffee mentioned the share of all units represented by multifamily units, but these units contributed to greater energy inefficiency for the total housing market because new multifamily units had more bedrooms and were larger. Space per person increased even more quickly because rental households, like owner households, became significantly smaller on the average in the 1970s. Use of space was expansive throughout the market. An important question is whether this trend will turn around in the 1980s. I do not know. Both builders and building materials manufacturers are concerned about it.

The General Electric Corporation is seeking to sponsor an extensive study of the potential demand for housing in the 1980s in relation to structural characteristics and components. GE has already approached the Joint Center at Harvard and MIT about being the umbrella group for such a study. GE is concerned lest it be in a position similar to that of the U.S. automobile industry if there is a quick shift of demand to small Japanese appliances, especially since GE's capital investment is quite large.

Builders are definitely beginning to adapt to higher energy costs. There is a clear inclination among the leaders of the NAHB to adapt to the realities of the 1980s as embodied in higher housing costs, high energy costs, and significant changes in the entire housing market in general.

The average size of households having grown smaller, it becomes difficult to know what the overall response to energy costs will be. How will demographics affect the amount of space demanded? At some point, the sheer number of smaller households must increase the demand for space unless households begin to double up. That brings me to an important point.

An understanding in more detail of what happened to the use of the existing housing stock in the 1970s is needed. In 1980, the Census Bureau counted 3.4 million more housing units than it had expected. Where those units came from must somehow be explained. Some portion can be explained as the result of an undercount in the 1970s. Another explanatory factor may be conversions. A much greater than expected number of added units may have been converted from the existing stock and from nonresidential structures. But in any event, it is essential to understand the components of change in the inventory much better than they are understood now, because similar changes will have a great effect during the 1980s.

This issue is also related to the occurrence of future urban concentrations. I have a hypothesis that I have begun to see confirmed by my discussions with builders. The greatest urban concentrations of the 1980s may be in and around the new central business districts (CBDs). These new CBDs will be located in those areas of suburban rings that are ideally situated to hold greater urban residential densities. An example can be observed here in the Washington, D.C., area in the massive new office and retail development in Tyson's Corner, Virginia, about fifteen miles from downtown Washington. It also contains growing clusters of multifamily homes and town houses. I suspect that this will also be happening elsewhere in the country. Such a development may change the whole mix between single-family and multifamily structures in the 1980s.

A continuing increase in the average size of multifamily units may also be seen, because they are more likely to be owner-occupied. In contrast, the average size of a unit in the single-family sector may decline with the movement toward town houses and zero-lot-line homes. Detached units will also be on smaller lots because of rising land costs.

In summary, Jaffee's paper deals with what happened in the 1970s. It shows that consumers were still space-expansive. There is no clear evidence that that tendency has yet changed. But what of the future? How will the new households yet to be formed and the first-time home buyers respond to higher costs and higher capital costs in the 1980s? I do not know, but the home-building industry will soon have to guess at the answers.

ROGER W. SCHMENNER

Energy and the Location of Industry

THE huge increases in energy costs and the difficulties of assuring uninterrupted supplies of oil and natural gas came as rude shocks to much of manufacturing America during the 1970s. During the last decade, a good deal of industrial thought has been devoted to the use and conservation of energy, to the point that many companies can now be said to pursue well-defined energy plans. Such plans are often far-ranging, so it is natural to ask to what degree location patterns of industry have been altered in accordance with companies' energy plans.

The question can be better stated if it is noted that energy has two general functions in most manufacturing: as input to the production process itself and as fuel for the transport of supplies to the factory or of outputs from it. Of course, one of these functions may be vastly more important to the company than the other and thus more likely to influence location patterns. Both have long been recognized, however. The costs of process energy have historically been of critical importance to such industries as aluminum smelting, and proximity to the market or to suppliers has been important to many other industries. Such industries could well be expected to keep energy considerations high on their lists of things that influence location. More interesting are those industries that have only recently become concerned about energy and the choice of location; these industries are the ones whose change of location is most apt to be visible.

Coping with Higher Real Costs of Energy

The demand for energy, like that for any other factor of production, is never perfectly inelastic, and the rapid rise of energy costs in the

1970s therefore touched off considerable substitution of other factors for energy. American industry in 1977 actually used fewer British thermal units of energy for heat and power (12,929 trillion) than it did in 1971 (13,008.3 trillion). During that same period, the cost of the energy used more than tripled.[1] Industry has clearly been substituting other factors of production for energy.[2] Such substitution has been the most visible feature of the effect of energy on manufacturing, and it has taken a variety of forms. The most dramatic kind of substitution has involved technological changes in production processes or modes of transportation. Less showy, although still effective, substitutions have not involved advances in hardware so much as procedural changes in the use of energy. Still other substitutions have been of less expensive or more reliable sources of energy for expensive or unreliable sources. Each of these forms of substitution merits some discussion.

Technological Adaptation

In all but a handful of industries—generally those that either use or manufacture heavy commodities—the most important mode of transportation by far is trucking. And, much as has the automotive industry, the trucking industry has been swept by technological changes that have saved millions of gallons of fuel. Some of the changes have been dramatic.[3] Of the heavy trucks on the road, 95 percent have diesel engines, an increase from 87 percent in 1973. More important, 65 percent of the present fleet of heavy trucks are equipped with fuel-economy diesels, whereas only 15 percent were in 1973. Of even the light-truck fleet, 10 percent are diesels, an increase from less than 2 percent in 1973. A number of other changes took place during the same period in the important heavy-truck fleet: the prevalence of demand-actuated fan clutches grew from use by only 4 percent of the fleet to use by 83

1. See U.S. Bureau of the Census, *Census of Manufactures, 1977: Fuels and Electric Energy Consumed*, pt. 1, *Industry Groups and Industries* (rev.), MC77-SR4 (Government Printing Office, 1980), table 2.

2. Of course, patterns of demand and the state of the national economy can substantially affect energy use. Some industries, among which primary metals is outstanding, have suffered declines in their utilization of capacity since the early 1970s, thereby reluctantly contributing to national conservation of energy.

3. The following figures are found in the section on energy efficiency in American Trucking Associations, "American Trucking Trends 1979–80" (Washington, D.C.: ATA, 1981). The most recent data are for 1979.

percent; radial tires now roll on 34 percent of the heavy trucks, whereas only 3 percent were so equipped in 1973; and wind deflectors operate on 13 percent of the fleet, an increase from less than 1 percent in 1973. These advances, it is estimated, saved more than 1.7 billion gallons of fuel in 1979, or about 10 percent of all the fuel used by trucks in that year.

The success stories of new energy-saving technological advances in production processes are legion. Some of the advances are general and have been applied in many industries—more fuel-efficient boilers; waste-heat-recovery systems; gas-recovery systems; better instrument monitoring and control of heating, cooling, and ventilation; more energy-efficient lighting. Others, however, are specific to the production processes themselves: wet grinding of phosphate rock and the use of cross-pipe reactors in the manufacture of fertilizers; continuous casting of steel strip and new cathodes for aluminum reduction in metallurgy; improved heat-exchange equipment in oil refining; dry-process rather than wet-process manufacture of cement; new furnace designs and improved controls in glassmaking; the use of bark dryers and the removal of excess air in drying ovens in pulp- and papermaking.[4] The future promises to be even better; advances in coal-based chemistry, a new means of aluminum reduction, hot boning of meat, and dry milling of corn, to name but a few, are energy-saving processes likely to be adopted widely in the near future.

In many instances, old technology rather than new has successfully been brought to bear in a variety of ways to reduce the dependence of factories on energy in conventional forms. Among these process-related changes are the use of otherwise waste heat (steam) to operate generators of electricity—cogeneration—mainly at complexes that manufacture forest products, primary metals, or chemicals; insulation of buildings and machinery; construction of larger production facilities to reduce energy costs as a percentage of sales—since larger blast furnaces, for example, save coke, larger aluminum plants save electricity, and longer cement kilns save fuel; the creation of worker awareness programs to promote conservation of energy; and the choice of interruptible power from utilities, saving the high cost of peak power generation.

Another category of energy saving involves redesign and even switch-

4. Most of these examples are derived from U.S. Department of Energy, *The Industrial Energy Efficiency Improvement Program*, DOE/CE-0015 (National Technical Information Service, 1980).

Table 1. *Consumption of Energy in 1972 and 1979, by Type of Fuel, All Manufacturing Corporations That Reported*[a]

Type of fuel	Consumption (billions of Btu) 1979	1972	Percent change, 1972–79
Electricity	3,000,188.34	2,589,537.51	15.86
Natural gas	5,229,158.07	6,069,551.11	−13.85
Propane	35,179.67	43,587.33	−19.29
LPG	43,287.72	38,118.07	13.56
Bituminous coal	2,687,453.44	2,835,280.03	−5.21
Anthracite coal	21,375.08	25,172.27	−15.08
Coke	276,185.00	213,817.19	29.17
Gasoline	38,085.33	18,572.04	105.07
Distillate fuel oil	223,868.47	228,013.83	−1.82
Residual fuel oil	1,352,227.00	1,419,169.00	−4.72
Petroleum coke	518,000.62	514,067.25	0.77
Purchased steam	216,992.87	229,189.25	−5.32
Other	1,819,922.22	1,593,810.60	14.19
Total energy consumed	15,461,923.83	15,817,885.48	−2.25

Source: U.S. Department of Energy, *The Industrial Energy Efficiency Improvement Program*, DOE/CE-0015 (National Technical Information Service, 1980), p. 65.

a. The Btu consumption given here is higher than that noted earlier, the source of which was the 1977 Census of Manufactures. The discrepancy may be attributable to restriction of the earlier figures to purchased energy. Energy produced and consumed in the same establishment—coke-oven gas, for example—and fuels used as raw materials, such as fuel oil and natural gas for carbon black, are excluded from those figures.

ing of products. The most dramatic examples of product redesign are automobiles, which have been downsized to reduce weight, thereby saving both gasoline for the consumer and process energy for the producer, and reconfigured to use gasoline more efficiently.

Substituting among Types of Energy

It has been national policy since the early 1970s to rely less on fossil fuels, such as oil and natural gas. In large measure this goal has been realized (see table 1), even as manufacturing output during the same period rose 17 percent. As reported by the U.S. Department of Energy, the mix of fuels consumed by industry has shifted since 1972. Less oil and natural gas are being used and other fuels are being substituted for them. Electricity is seen as more reliably supplied than either oil or natural gas. The same table shows a decline in the use of coal, but that is largely explained by a 9 percent drop in the output of the primary metals industry, mainly steel, which is the principal user of coal in

Table 2. *Energy Intensity of Major Industries*
Millions of dollars unless otherwise specified

SIC code	*Industry*	*Energy costs*	*Value added*	*Energy costs as percent of value added*	*Rank*
20	Food processing	2,537.8	56,062.2	4.5	8
21	Tobacco	63.9	4,334.1	1.5	17
22	Textiles	1,136.6	16,104.5	7.1	6
23	Apparel	285.8	19,671.3	1.5	18
24	Lumber	726.5	16,222.9	4.5	9
25	Furniture	198.3	8,921.7	2.2	11
26	Paper	2,960.4	22,170.9	13.4	4
27	Printing	421.7	31,979.5	1.3	20
28	Chemicals	6,448.7	56,720.6	11.4	5
29	Petroleum	2,445.4	16,377.6	14.9	2
30	Rubber and plastics	962.4	19,740.0	4.9	7
31	Leather	82.6	3,718.8	2.2	12
32	Stone, clay, glass, and cement	2,587.4	19,129.6	13.5	3
33	Primary metals	7,043.6	37,568.2	18.8	1
34	Fabricated metals	1,348.7	45,511.5	3.0	10
35	Machinery	1,268.0	67,222.5	1.9	16
36	Electrical and electronic machinery	986.7	50,365.7	2.0	14
37	Transportation equipment	1,420.6	64,291.1	2.2	13
38	Instruments	272.5	18,762.0	1.5	19
39	Miscellaneous manufacturing	199.9	10,291.0	1.9	15

Sources: U.S. Bureau of the Census, *Census of Manufactures, 1977: General Summary*, MC77-SR1 (Government Printing Office, 1981), pp. 30–44; and Bureau of the Census, *Census of Manufactures, 1977: Fuels and Electric Energy Consumed*, pt. 1, *Industry Groups and Industries* (rev.), MC77-SR4 (GPO, 1980), p. 4-14.

manufacturing. The table also shows that, while gasoline is not an important industrial fuel, its use has increased dramatically since 1972. This increase, apparently, is not easily explained.

Much fuel substitution has come about through the building of new boilers or modification of existing ones so that different fuel is burned—coal, say, instead of oil—or so that more than one fuel can be burned. In production processes that require heat, such as those in which kilns are used, electricity has been substituted for natural gas to some extent.

Differences in Energy Intensity among Industries

The incentive of any industry to use proportionately less energy is naturally dependent on the intensity with which energy is used in the production process. Tables 2 and 3 document one measure of industrial

Table 3. *The Twenty-five Most Energy-Intensive Industries: Three-Digit SIC Industries Ranked by Energy Costs as a Percent of Value Added, 1977*

Rank	*SIC code*	*Industry*	*Energy costs as percent of value added*
1	324	Cement	43.8
2	266	Building paper (including asphalt, tar, and asbestos)	33.6
3	263	Paperboard	32.1
4	281	Industrial inorganic chemicals (chlorine, industrial gases)	29.0
5	333	Primary smelting and refining of nonferrous metals (aluminum)	27.3
6	331	Blast furnaces and steelworks	24.6
7	262	Paper mills	23.9
8	261	Pulp mills	21.8
9	325	Structural clay products (bricks, refractories)	21.4
10	286	Industrial organic chemicals	17.5
11	287	Agricultural chemicals (fertilizers)	16.8
12	291	Petroleum refining	15.6
13	322	Glass and glassware (including bottles)	14.6
14	321	Flat glass	14.6
15	207	Fats and oils	14.3
16	226	Dyeing and finishing textiles	14.1
17	303	Reclaimed rubber	13.3
18	282	Plastic materials and synthetic resins and rubber	13.1
19	295	Paving and roofing material	13.0
20	332	Iron and steel foundries	12.0
21	334	Secondary smelting and refining of nonferrous metals	11.7
22	339	Miscellaneous primary metal products (heat treating)	11.2
23	249	Miscellaneous wood products (particle board)	10.3
24	214	Tobacco stemming and redrying	9.2
25	329	Miscellaneous nonmetallic mineral products	8.7

Sources: Bureau of the Census, *Census of Manufactures, 1977: General Summary*, pp. 30–44; and Bureau of the Census, *Industry Groups and Industries*, pp. 4-14–4-30.

energy intensity for industries having two-digit and three-digit Standard Industrial Classification codes. It is clear from these tables that the incentive to conserve or to switch forms of energy is great for the heavy, commodities-producing industries that use huge quantities of it. For many other industries, energy costs are of little consequence.

The apparent consequence, or inconsequence, of energy for the location of industry depends, naturally, on what energy costs are compared with. In tables 2 and 3, energy costs are compared with value added, largely because the data are conveniently available. They could as well have been compared with total value of shipments or some long-term measure of profitability. The former comparison would diminish the apparent importance of energy while the latter would heighten it. Other comparisons could doubtless be made. No one measure is clearly preferable, since what matters at the company level for the location of an industry is how geographic variations in energy costs compare to geographic variations in the costs of other factors of production and in sales. The factors involved in such comparisons, and the variations to which they are subject, are typically idiosyncratic to the company, so general measures of energy intensity, such as those in tables 2 and 3, are, at best, suggestive rather than definitive.

Geographic Disparities

There are some real differences in the prices of various forms of energy, as documented in tables 4–6. The cost of electricity, by state and region, is examined in table 4. The lowest electric rates are found in those states of the Pacific Northwest—Washington, Oregon, Montana, Idaho, Wyoming—where hydroelectric power is plentiful. The highest rates are now found in Hawaii and several states of the Northeast, such as Massachusetts, Rhode Island, New York, and New Jersey. In some instances there are marked differences in the average cents per kilowatt hour paid by industry; average rates in Washington are equal to only 15 percent of those in Hawaii, and rates in Oregon are equal to only 31 percent of those in Massachusetts or Rhode Island. For the most part, however, rates for electricity are fairly close to one another. Average 1979 rates in thirty-two of the fifty states and the District of Columbia were within 20 percent of the national average. The closeness of the state rates, moreover, has become more pronounced during at least the past twenty years. As table 4 makes clear, the standard deviations of the index numbers that document cross-state disparities have been declining steadily. This suggests that while there are still sizable dollar advantages to be gained on electric rates by locating in one state rather than another, in relative terms those advantages are declining.

Data on the costs of so-called residual oil, which is a grade of fuel oil frequently used by industry, are presented in table 5. The pricing of residual oil is seen to be more nearly uniform that that of electric power. Residual oil is likely to be similarly priced in all the states of a region. And, as is true of electric power, the variation in price from one state to another has been declining consistently during at least the past twenty years.

Natural gas prices are compared in the same way in table 6. As can be seen there, the greatest diversity in prices among states is in natural gas. The Northeast and the Pacific Northwest understandably suffer the most from being far removed from the largest deposits of natural gas. The Southwest and the Plains states enjoy the lowest prices. But, as is true of the prices of electricity and residual oil, the distribution of natural gas prices among states is steadily becoming tighter, and variations have been consistently decreasing.

In tables 7–9 the prices of a composite measure of energy use, the kilowatt hour (kwh) equivalent, are compared. This summary statistic is then used to assess changes in energy prices among both states and industries. The weights applied to the prevailing fuel prices in each state or industry were derived from the Btu content of each fuel and its actual relative use in each state or industry. During the twenty years 1958–77, the nationwide average price per kwh approximately doubled. Individual state prices changed much more wildly, although the rank order of states stayed roughly the same. Nevertheless, some interesting trends were to be observed: the Mountain states improved their already strong relative showing, with energy prices actually lower in states such as Wyoming, Montana, Utah, and Colorado than in Texas and Oklahoma, and the ranks of states as diverse as Alaska, Utah, Idaho, Maine, Virginia, and Wisconsin improved at least ten places, while Alabama, Kentucky, Mississippi, New Mexico, the District of Columbia, Georgia, and especially California, dropped at least ten places in rank between 1958 and 1977.

Table 7 reinforces the finding that energy prices seem to be converging. There is notably less variation in 1977 than in 1958.

In table 8 both nominal and real energy costs paid by various large industries are examined. As might be expected, for all the heavily energy-intensive industries—primary metals; stone, clay, glass, and concrete; petroleum; chemicals; and paper—energy costs per kwh equivalent are lower than average. This is in part because of the mix of fuels used and purchasing arrangements, in part because of the technology selected,

Table 4. *Costs of Industrial Electricity, by Region and State, Selected Years, 1960–79*
Index (United States = 1.00)

Region and state	1979	1975	1970	1965	1960
New England					
Connecticut	1.35	1.56	1.31	1.43	1.54
Maine	1.03	1.09	1.21	1.34	1.34
Massachusetts	1.50	2.02	1.53	1.66	1.68
New Hampshire	1.36	1.45	1.33	1.47	1.51
Rhode Island	1.50	1.65	1.41	1.56	1.64
Vermont	0.96	1.17	1.34	1.42	1.44
Mid Atlantic					
New Jersey	1.46	1.59	1.28	1.26	1.33
New York	1.48	1.63	1.38	1.33	1.29
Pennsylvania	1.09	1.18	1.04	1.05	1.14
East North Central					
Illinois	1.10	1.08	1.21	1.24	1.33
Indiana	0.98	0.84	1.03	1.10	1.19
Michigan	0.71	1.18	1.16	1.14	1.25
Ohio	0.91	0.89	0.91	0.81	0.73
Wisconsin	1.00	1.06	1.29	1.29	1.34
West North Central					
Iowa	1.06	1.05	1.30	1.39	1.43
Kansas	1.01	0.91	1.15	1.19	1.10
Minnesota	0.96	1.02	1.27	1.41	1.54
Missouri	1.05	1.01	1.26	1.28	1.36
Nebraska	0.83	0.82	1.07	1.16	1.13
North Dakota	0.95	1.07	1.63	1.96	2.40
South Dakota	1.02	1.12	1.74	1.79	1.75
South Atlantic					
Delaware	1.26	1.35	0.94	0.90	1.24
District of Columbia	1.16	1.33	1.23	1.24	1.32
Florida	1.20	1.30	1.26	1.42	1.62
Georgia	1.03	1.16	1.00	1.02	1.06
Maryland	1.16	1.33	1.23	1.24	1.32
North Carolina	0.87	0.94	0.83	0.86	0.87
South Carolina	0.84	0.87	0.75	0.76	0.76
Virginia	1.14	1.20	1.04	1.05	1.13
West Virginia	0.79	0.96	0.77	0.77	0.79
East South Central					
Alabama	0.92	0.81	0.68	0.63	0.64
Kentucky	0.71	0.61	0.64	0.52	0.44
Mississippi	1.05	0.95	0.94	0.94	1.06
Tennessee	0.84	0.67	0.55	0.44	0.41
West South Central					
Arkansas	0.87	0.87	0.95	0.96	0.99
Louisiana	0.73	0.67	0.87	1.00	1.01
Oklahoma	0.85	0.73	1.09	1.17	1.20
Texas	0.92	0.76	0.86	0.98	1.04

Table 4 *(continued)*

Region and state	1979	1975	1970	1965	1960
Mountain					
Arizona	1.20	1.15	1.12	1.19	1.04
Colorado	0.94	0.95	1.27	1.37	1.33
Idaho	0.51	0.47	0.64	0.65	0.63
Montana	0.36	0.40	0.51	0.55	0.53
Nevada	0.93	0.87	0.90	0.72	0.66
New Mexico	1.23	0.92	1.23	1.30	1.32
Utah	0.95	0.84	1.14	1.21	1.13
Wyoming	0.55	0.56	1.00	0.99	1.18
Pacific					
Alaska	1.21	1.25	2.15	2.50	2.94
California	1.17	1.07	1.02	1.06	1.12
Hawaii	1.55	1.54	1.50	1.55	1.58
Oregon	0.46	0.43	0.51	0.52	0.52
Washington	0.24	0.29	0.38	0.39	0.39
Addenda					
All United States (cents per kwh)	3.43	2.43	1.32	1.29	1.34
Standard deviation of state distribution (index units)	0.28	0.35	0.33	0.40	0.46

Source: Edison Electric Institute, *Statistical Yearbook of the Electric Utility Industry*, 1960, 1965, 1970, 1975, and 1979 (Washington, D.C.: EEI), tables 22 and 36.

and in part because of location chosen. The industries that are not energy-intensive are content to pay higher costs per kwh equivalent.

In table 9 costs per kwh equivalent paid by the five most energy-intensive industries in ten representative states are compared. In general the variation in energy costs among industries in any one state is considerably less than the variation in costs among states for any one industry. This suggests that many of the largest industries have chosen their locations to take advantage of differences in energy costs. The assumption that production technology matches the availability of resources is reinforced by the observations that the lowest energy costs in the state of Washington are enjoyed by primary metals, chiefly aluminum, using hydroelectric power, that the lowest costs in Texas are enjoyed by petroleum refining, and that the lowest costs in tree-rich states such as Alabama, Massachusetts, North Carolina, and Ohio are enjoyed by the paper industry.

The Location Decision Process

Whether any of the geographic disparities noted in the preceding section have any measurable influence on the changing geographic pattern

Table 5. *Costs of Residual Oil, by Region and State, Selected Years, 1960–77*
Index (United States = 1.00)

Region and state	1977	1975	1970	1965	1960
New England					
Connecticut	1.10	0.98	0.75	0.84	0.83
Maine	0.99	0.91	0.82	0.85	0.87
Massachusetts	1.07	0.94	0.73	0.84	0.87
New Hampshire	1.04	0.93	0.76	0.84	0.86
Rhode Island	1.00	0.95	0.76	0.85	0.87
Vermont	1.04	0.93	0.76	0.84	0.86
Mid Atlantic					
New Jersey	1.02	0.92	0.93	0.93	0.86
New York	1.91	0.91	0.99	1.01	0.90
Pennsylvania	1.03	0.93	0.96	0.84	0.82
East North Central					
Illinois	1.13	0.98	1.15	1.06	1.02
Indiana	1.13	0.97	1.13	1.11	1.06
Michigan	1.13	0.95	1.10	1.16	1.11
Ohio	1.13	0.97	1.13	1.11	1.06
Wisconsin	1.13	0.97	1.13	1.11	1.06
West North Central					
Iowa	1.02	1.05	1.06	1.07	1.02
Kansas	1.02	1.05	1.06	1.07	1.02
Minnesota	1.08	1.05	1.06	1.07	1.02
Missouri	0.80	1.05	1.06	1.07	1.02
Nebraska	1.02	1.05	1.06	1.07	1.02
North Dakota	1.02	1.05	1.06	1.07	1.02
South Dakota	1.02	1.05	1.06	1.07	1.02
South Atlantic					
Delaware	0.97	0.85	0.79	0.63	0.59
District of Columbia	0.97	0.85	0.79	0.63	0.59
Florida	1.01	0.84	0.79	0.75	0.69
Georgia	0.96	0.83	0.79	0.75	0.69
Maryland	1.00	0.94	0.79	0.74	0.67
North Carolina	0.95	0.84	0.79	0.63	0.59
South Carolina	0.96	0.84	0.79	0.75	0.72
Virginia	0.97	0.83	0.77	0.75	0.73
West Virginia	0.97	0.84	0.79	0.63	0.59
East South Central					
Alabama	0.81	0.78	0.82	0.80	0.69
Kentucky	0.82	0.78	0.82	0.80	0.69
Mississippi	0.82	0.78	0.82	0.80	0.69
Tennessee	0.81	0.77	0.82	0.80	0.69
West South Central					
Arkansas	0.86	0.81	0.88	0.83	0.73
Louisiana	0.86	0.95	0.82	0.83	0.74
Oklahoma	0.85	0.80	0.76	0.75	0.61
Texas	0.87	0.82	0.82	0.80	0.69

Table 5 *(continued)*

Region and state	1977	1975	1970	1965	1960
Mountain					
Arizona	0.95	1.11	1.25	1.22	1.18
Colorado	0.94	1.11	1.25	1.22	1.18
Idaho	0.94	1.11	1.25	1.22	1.18
Montana	0.92	1.11	1.25	1.22	1.18
Nevada	0.94	1.11	1.25	1.22	1.18
New Mexico	0.95	1.11	1.25	1.22	1.18
Utah	0.95	1.11	1.25	1.22	1.18
Wyoming	0.94	1.11	1.25	1.22	1.18
Pacific					
Alaska	1.02	1.14	1.27	1.23	1.20
California	1.07	1.17	1.27	1.23	1.20
Hawaii	1.06	1.16	1.27	1.23	1.20
Oregon	1.02	1.11	1.27	1.23	1.20
Washington	1.07	1.16	1.27	1.23	1.20
Addenda					
All United States (dollars per barrel)	14.40	14.17	4.00	3.46	3.86
Standard deviation of state distribution (index units)	0.09	0.12	0.20	0.20	0.21

Source: Department of Energy, Energy Information Administration, *State Energy Fuel Prices by Major Economic Sector from 1960 through 1977* (GPO, 1979), table B-1, pp. 83–84.

of industry in America depends significantly on the nature of the location decision process within the company. It is worth while to place the location decision process in perspective before attempting to assess the influence of energy in actual choices.

The location decision, to my mind, is properly regarded as part of a series of decisions that, taken together, determine the composition and character of the manufacturing capacity of a company. The first requirement is a firm grasp of the reasons a company may need new capacity, and what that new capacity should do for it, before the company plunges into the selection of an appropriate site. Energy and the geographic disparities associated with it exert broader influences than merely those associated with the selection of particular sites. As noted earlier, companies have found a number of ways of coping with higher real energy costs. It may well be that in response to a hefty increase in the real price of energy a company would be better off to expand in its present location than to place any new capacity in a new location in which energy was lower in cost. The savings in the cost of energy derived by expanding on site might dwarf any savings attributable to an alternate location. For this reason, in some industries at least, higher real energy prices could

Table 6. *Costs of Natural Gas, by Region and State, 1970, 1975, and 1979*
Index (United States = 1.00)

Region and state	1979	1975	1970
New England			
Connecticut	1.54	2.29	2.61
Maine	1.60	2.12	3.71
Massachusetts	1.42	2.31	2.63
New Hampshire	1.46	1.81	2.66
Rhode Island	1.21	2.15	1.87
Vermont	1.57	1.11	1.26
Mid-Atlantic			
New Jersey	1.36	1.63	1.68
New York	1.29	1.44	1.97
Pennsylvania	1.12	1.42	1.50
East North Central			
Illinois	1.13	1.12	1.21
Indiana	0.96	0.92	1.24
Michigan	1.04	1.24	1.42
Ohio	1.04	1.13	1.42
Wisconsin	1.07	1.06	1.39
West North Central			
Iowa	0.92	0.79	0.95
Kansas	0.76	0.60	0.68
Minnesota	0.93	0.87	1.05
Missouri	0.82	0.75	0.95
Nebraska	0.75	0.68	0.84
North Dakota	1.19	1.19	1.16
South Dakota	0.89	0.66	0.87
South Atlantic			
Delaware	1.44	1.21	1.39
District of Columbia	1.38	1.51	2.13
Florida	0.88	0.80	1.00
Georgia	0.96	0.82	0.95
Maryland	1.09	1.37	1.79
North Carolina	1.15	1.33	1.29
South Carolina	0.88	0.86	1.13
Virginia	1.06	1.13	1.34
West Virginia	1.06	1.06	1.24
East South Central			
Alabama	0.96	0.78	0.84
Kentucky	0.91	0.87	1.24
Mississippi	0.94	0.80	0.74
Tennessee	0.93	0.82	0.95
West South Central			
Arkansas	0.70	0.68	0.71
Louisiana	0.86	0.75	0.61
Oklahoma	0.75	0.67	0.71
Texas	1.04	1.06	0.61

Table 6 *(continued)*

Region and state	1979	1975	1970
Mountain			
Arizona	0.94	0.86	1.00
Colorado	0.86	0.63	0.68
Idaho	1.11	1.16	1.13
Montana	1.00	0.93	0.89
Nevada	1.05	1.07	1.08
New Mexico	0.90	0.69	0.74
Utah	0.71	0.65	0.74
Wyoming	0.74	0.58	0.63
Pacific			
Alaska	0.40	0.52	1.37
California	1.09	1.05	0.97
Hawaii	3.95	6.41	5.76
Oregon	1.31	1.42	1.08
Washington	1.24	1.27	1.00
Addenda			
All United States (dollars per million Btu)	2.24	0.99	0.38
Standard deviation of state distribution (index units)	0.476	0.866	0.878

Source: American Gas Association, Department of Statistics, *Gas Facts, 1979* (Arlington, Va.; AGA, 1980), pp. 112, 113, 119.

have a significant influence on the location of plants merely by helping to forestall the start-up of any new plants, thereby fixing the geographic configuration of manufacturing facilities.

This view is backed in part by some empirical findings for selected industries. In response to a survey of industry trade associations conducted for this paper, three of nineteen industries, notably chemicals, indicated that larger plants were at an advantage in curbing the effects of higher energy prices. For most industries, however, the size of the plant has little or nothing to do with the importance of energy to its operation; on-site expansion, while popular on other counts, is not always to be preferred to the opening of new plants because of energy considerations.

It should also be noted that changes in location, whatever the reasons for them, are unlikely to take place quickly. Given the place of the location decision in the even grander process of determining on-site expansion, relocation, or the opening of a new plant, it is not surprising that the ability of a company to alter its energy costs and consumption through a change of location is likely to take some time. It takes time to plan new locations and time to determine whether particular plants can

Table 7. *Index of Costs of All Fuels per Kilowatt Hour Equivalent, by Region and State, 1958 and 1977*
Index (United States = 1.00)

Region and state	1958	1977
New England		
Connecticut	3.93	1.63
Maine	7.15	1.05
Massachusetts	3.56	1.54
New Hampshire	6.26	1.29
Rhode Island	3.34	1.59
Vermont	5.89	1.56
Mid Atlantic		
New Jersey	3.46	1.34
New York	2.87	1.22
Pennsylvania	1.45	1.13
East North Central		
Illinois	1.63	1.11
Indiana	1.57	0.99
Michigan	2.21	1.16
Ohio	1.41	1.10
Wisconsin	3.20	1.05
West North Central		
Iowa	1.06	0.88
Kansas	0.40	0.80
Minnesota	1.45	0.99
Missouri	1.17	0.95
Nebraska	0.79	0.78
North Dakota	1.42	0.69
South Dakota	0.90	0.94
South Atlantic		
Delaware	2.86	1.29
District of Columbia	2.98	1.99
Florida	1.76	1.12
Georgia	1.05	1.11
Maryland	3.07	1.13
North Carolina	2.54	1.22
South Carolina	1.66	1.08
Virginia	3.17	1.09
West Virginia	1.22	0.86
East South Central		
Alabama	0.86	1.04
Kentucky	1.02	1.20
Mississippi	0.48	0.87
Tennessee	0.97	1.02
West South Central		
Arkansas	0.32	0.84
Louisiana	0.20	0.52
Oklahoma	0.28	0.77
Texas	0.21	0.77

Table 7 *(continued)*

Region and state	*1958*	*1977*
Mountain		
Arizona	0.74	0.87
Colorado	0.58	0.69
Idaho	1.38	0.84
Montana	0.75	0.65
Nevada	1.02	0.78
New Mexico	0.63	0.90
Utah	0.99	0.68
Wyoming	0.51	0.54
Pacific		
Alaska	20.91	0.90
California	0.76	1.24
Hawaii	9.22	1.29
Oregon	1.31	0.89
Washington	0.76	0.70

Sources: For 1958, Bureau of the Census, *Census of Manufactures, 1958*, vol. 1, *Summary Statistics* (GPO, 1961), pp. 6-26–6-29; and for 1977, *Census of Manufactures, 1977*, vol. 1, *Subject Series* (GPO, 1981), pp. 4-60–4-141.

be salvaged and their closings avoided. There are frequently energy-saving investments a company can make, moreover, that attenuate the effects of escalations in energy costs at existing facilities. To make such energy-saving investments is typically much less expensive than to shift production to an entirely new site.

Once it has been deemed desirable to relocate, however, how do energy considerations enter into the selection process? For almost all corporations of any consequence, the evaluation of potential plant locations includes systematic consideration of costs and benefits. As much as possible will be quantified—site and site-preparation costs, the costs of construction or purchase and renovation, the costs of equipment and of labor and fringe benefits, start-up costs, such as training and initial inefficiencies, working capital requirements, the expenses of freight in and out, energy use, taxes of all kinds, the expense of relocation of personnel, and expected revenues. Still, not everything that is important can be quantified; the decision is usually too complex to be evaluated solely by the numbers.

In addition, the complexity of the decision often inhibits the generation of potential sites. In order to simplify the choice and to enable companies to uncover acceptable sites readily through the traditional site-generation channels, most companies depend on identification of one or two controlling concerns, by means of which they can screen regions, states,

Table 8. *Costs of All Fuels, by Industry, 1958 and 1977*
Cents per kilowatt hour equivalent

Industry	Cost		Inflation-adjusted cost (1967 cents)	
	1958	1977	1958	1977
All industries	0.21	0.88	0.32	0.63
Food	0.23	0.91	0.35	0.65
Tobacco	0.24	1.05	0.36	0.75
Textiles	0.30	1.14	0.46	0.82
Apparel	0.52	1.49	0.78	1.06
Lumber	0.45	1.09	0.69	0.78
Furniture	0.34	1.28	0.52	0.92
Paper	0.16	0.77	0.25	0.55
Printing	0.61	1.56	0.93	1.12
Chemicals	0.17	0.74	0.26	0.53
Petroleum	0.10	0.64	0.15	0.46
Rubber and plastics	0.27	1.21	0.41	0.86
Leather	0.30	1.22	0.46	0.87
Stone, clay, and glass	0.17	0.71	0.26	0.50
Primary metals	0.21	0.95	0.33	0.68
Fabricated metals	0.36	1.16	0.55	0.83
Machinery, except electrical	0.34	1.27	0.52	0.91
Electrical machinery	0.37	1.35	0.56	0.97
Transportation equipment	0.33	1.23	0.49	0.88
Instruments	0.34	1.19	0.51	0.85
Miscellaneous manufacturing	0.38	1.39	0.58	0.99

Sources: Bureau of the Census, *Census of Manufactures, 1958*, vol. 1, p. 6–8; and ibid., 1977, vol. 1, p. 4-14.

and often metropolitan areas and can provide guidelines for those working to generate potential sites. The principal controlling concerns brought to my attention from interviews and survey data, in no particular order, include:[5]

Labor costs. Labor costs are a primary constraint on highly competitive industries, such as apparel, leather, furniture, and consumer electronics, whereas they are of relatively little concern to many capital-intensive operations.

Unionization of labor. Employing a new work force that is nearly impossible to organize is perhaps the most prized side benefit of a new

5. See Roger W. Schmenner, *Making Business Location Decisions* (Prentice-Hall, 1982), especially chap. 3.

Table 9. *Costs of All Fuels in Representative States, by Industry, 1962 and 1977*
Cents per kilowatt hour equivalent

State	*Year*	*Paper*	*Chemicals*	*Petroleum*	*Stone, clay, glass*	*Primary metals*
Alabama	1962	1.74	1.41	1.67	0.82	2.08
	1977	0.93	1.18	. . .	0.94	1.06
California	1962	1.97	1.68	1.52	1.12	1.98
	1977	1.32	1.39	1.54	1.13	1.23
Colorado	1962	1.16	1.89	1.71	0.84	. . .
	1977	0.88	0.90	0.69	0.67	0.65
Massachusetts	1962	16.70	7.19	28.19	. . .	5.58
	1977	1.42	1.44	1.56	1.79	1.98
Minnesota	1962	2.82	5.13	3.91	1.90	2.05
	1977	1.08	1.01	1.19	1.29	1.17
North Carolina	1962	6.16	7.22	. . .	1.63	. . .
	1977	0.96	1.37	1.33	1.18	1.31
New Jersey	1962	14.53	8.63	25.73	2.94	6.51
	1977	1.52	1.51	1.33	1.48	1.41
Ohio	1962	8.62	2.64	2.90	1.78	2.46
	1977	0.89	1.28	1.50	1.06	1.03
Texas	1962	0.88	0.50	0.70	0.69	0.65
	1977	0.95	0.90	0.86	0.96	1.13
Washington	1962	1.57	1.89	4.52	. . .	1.10
	1977	0.82	0.82	1.17	0.82	0.46

Sources: Bureau of the Census, *Census of Manufactures, 1963*, vol. 1, *Summary and Subject Statistics* (GPO, 1966), pp. 7-38–7-64; and ibid., 1977, vol. 1, pp. 4-60–4-141.

plant site, and for many companies it becomes the controlling consideration. Such a concern ranks relatively high with companies that are sensitive to labor costs, a number of companies that make specialty products, and many makers of industrial equipment.

The quality of life in an area. The most fearsome competitive advantage that a high-technology company, especially, can wield is a happy and productive staff of engineers. Since proximity to markets or suppliers is not required by such companies, they can be extraordinarily footloose, locating in areas in which their engineers and managers find it attractive to live.

Proximity to markets. For the manufacturers of commodity or near-commodity items of a low value in relation to weight, a site near established markets is absolutely essential. Paper converting, printing, plastics fabrication, can making, and a number of other industries often find themselves constrained by transportation costs.

Proximity to supplies or resources. Conversely, several industries are tied to certain suppliers or resources. Paper mills must be near trees and water, fruit and vegetable processors near farms, petrochemical complexes near oil pipelines, and so on.

Proximity to other company facilities. There is a class of manufacturing plant that cannot live alone; it operates as a satellite plant to a base or mother plant. Without a steady stream of supplies, inventories of work in process, management expertise, engineering talent, and the like from a mother plant, such a plant would wilt under the stress.

From time to time, other considerations become controlling concerns of the management team charged with the location decision, but these are the primary ones, according to my research and that of others.[6] Teamed with these constraints are generally others that pertain to the site itself—it must have a railroad siding, must be on water, must have certain utilities, such as natural gas, and so on—but these are fairly low-level qualifications and can often be met within a fairly short radius. What is important to remember about any of these constraints on the location decision is that they are, by and large, outgrowths of the economics of the industry or of the company's particular manufacturing priorities. Only if a location passes successfully through one or two of these screens does a company then examine it on other, less constraining grounds.

Of the six primary or controlling concerns introduced above, the last three can be directly affected by the cost or availability of energy. Proximity to markets has traditionally been a primary factor in plant location decisions for a number of industries, and the recent rise in real energy costs has doubtless heightened the concern of management about such proximity. The same is true of proximity to supplies and natural resources. This concern, moreover, also includes location choices that take advantage of particular low-cost sources of energy for use in production processes themselves. The primary concern for proximity to other company plants generally has less to do with energy costs, since such a choice is more often dictated by time of managers and engineers in transit between plants than by the cost of moving goods or people between plants.

6. Ibid.

What is significant here is that these concerns were directing location choices before the real costs of energy rose to their present levels. At issue is whether the actual choices of location governed by these concerns could be improved upon. It may be that for plant locations governed by them, the recent increases in the real costs of energy have meant nothing dramatic for the geographic pattern of industry location, since the sites where energy is least expensive were already being picked. The influence of energy costs on plant location, while stronger then ever before, may not be noticeably different.

Of course, the cost and availability of energy may be a more important secondary influence than ever before on those plant location decisions that are not controlled by energy considerations. And, in those company situations where no controlling concern is readily identified—roughly a fourth to a third of all location choices—energy may have a powerful influence.

Assessing the degree to which energy prices and availability influence such decisions is extraordinarily difficult. A kind of Heisenberg Uncertainty Principle prevails in many of the statistical tests that are attempts to sort out the effects of such variables as energy prices from other influences on choice of location. The Heisenberg Uncertainty Principle of modern physics, loosely interpreted, suggests that the researcher's attempt to measure one aspect of the phenomenon in question may actually impede his ability to measure another aspect of the phenomenon at the same time. So it goes in isolating energy effects from other effects on location of industry. Trying to control for diverse energy prices in different areas leads the researcher to geographically based economic models in which the observations are defined for particular pieces of land, such as cities, counties, or SMSAs, that necessarily aggregate the decisions of individual companies and ruin the precision of many of the independent variables the observer might hope to include.[7] On the other hand, trying to control for variations in plant-specific influences on the choice leads the researcher to plant-based economic models, in which

7. In 1977, for example, I published a so-called evolutionary model of the firm's plant location decision in which tax rates had a definite and prescribed influence on location decisions. Yet when a geographically based economic model was used to isolate the effect of tax rates on location, using data generated by a simulation of the already known and well-prescribed evolutionary process, the tax-rate variable coefficient was both insignificantly different from zero and carried the unanticipated sign. Such a result calls into question the assumption that anything subtle about industry can be uncovered with the use of geographically based models. See Roger W. Schmenner, "Urban Industrial Location: An Evolutionary Model," *Journal of Regional Science*, vol. 17 (August 1977), pp. 179–94.

the observations are individual plants—a dummy variable, new or old plant, for example, as the dependent variable—models that cause him to ignore much of the geographic variation in energy prices that a company sorts through in evaluating regions, communities, and sites.

The Effects of Energy Prices and Availability on the Location of Industry: Uses of Process Energy

For the most part, solid research on energy used in production processes themselves as an influence on the location of industry has been scant. As a result circumstantial evidence has been the basis of some flawed reasoning. Some observers have noted that there remain regional differences in energy availability and prices and that corporate decision-makers have become more aware of energy considerations; putting these two observations together, they conclude that much of the growth of the South and Southwest is attributable to energy considerations. This conclusion does not follow. While energy considerations may be part of the explanation for the growth of the South and Southwest, there are a host of other considerations with much firmer claims on persuasiveness—labor rates, unionization, and market growth, for example. These observers ignore the technological advances that have liberated most factory locations from nearby sources of energy—electricity generation technology, interlocking electric power grids, and gas and oil pipelines, for example.

More sophisticated observers recognize that there are relatively few industries for which energy is something other than a minor influence on the plant-location decision.[8] Tables 2 and 3 are worth another look now. A classic energy-sensitive industry is aluminum reduction; 20–30 percent of the final price of the metal is accounted for by the cost of electric power. Certain chemicals industries—fertilizers, electrochemicals, and synthetic fibers—also demand significant amounts of energy. There are other energy-intensive industries, such as cement, steel, pulp and paper, and oil refining, but some of these—pulp and paper, refining—can be nearly self-sufficient in energy; others, such as steel and cement, depend for energy on coal, for which bulk transportation is well organized.

8. Henry L. Hunker, *Industrial Development: Concepts and Principles* (D. C. Heath, 1974); and Howard A. Stafford, *Principles of Industrial Facility Location* (Atlanta: Conway Publications, 1979).

These and other energy-intensive industries have frequently become less and less tied to cheap or readily available sources of energy as technology has spread energy resources out geographically.[9] Glassmaking was once tied to charcoal, then to sources of natural gas, but it is now more likely to be located near markets. Cement making, especially since the advent of the dry production process, is more footloose now than it was. Similar statements could be made of a number of industries.

Of concern in at least some quarters is the continued availability of energy, since it is recognized that the inability to procure energy is a more important deterrent to the location of a plant in a given area than the price of that energy. The unavailability of natural gas in some states during the mid 1970s was of concern to many companies and may have given impetus to a few new choices of location in areas such as the Southwest, where supplies are assured, and deterred new locations in gas-short areas. In addition, some electric utilities have had problems in overcoming the actions and legal maneuvers of environmental interest groups so that needed capacity for generating electricity, especially nuclear capacity, could be brought on line. These problems, and the delays associated with them, could deter some industries from locating new plants in the affected areas. The extent of any such deterrence has not yet been documented.

Political uncertainties about environmental regulations and their effects on energy matters may also deter new plant locations. For the most part, environmental regulations affect industrial location decisions independent of energy, but cases exist—conversion to coal-fired boilers, for example—in which energy and environmental matters have become intertwined. The effects of these uncertainties on location are not clear-cut. Often, such regulations stimulate on-site expansion of existing facilities at the expense of new plant openings, since it is usually easier to secure permits at existing sites, sometimes because trade-offs among sources of pollution can be arranged. More rural areas may also benefit somewhat at the expense of urban areas. The extent of the influence of these uncertainties remains unexplored, however.

I know of only two studies in which systematic efforts have been made to sort out the influence of process energy considerations on the location of industry. Dennis Carlton of the University of Chicago has investigated

9. For some interesting examples, see E. Willard Miller, *Manufacturing: A Study of Industrial Location* (Pennsylvania State University Press, 1977).

the locations of new single-establishment and branch plants in three narrowly defined industries:[10] plastic products (standard industrial classification [SIC] 3079), electronic transmitting equipment (SIC 3662), and electronic components (SIC 3679). Carlton's models in these papers are innovative and sophisticated, and the estimations, using metropolitan area observations, were accomplished in nonroutine ways. It will suffice here only to report his findings. Included as independent variables in Carlton's estimations were the prices of electricity and natural gas. In nearly all the estimations, these energy prices were significant influences on the location of new plants in at least two of the three industries studied. In fact, the importance of energy prices in relation to something such as wages is surprisingly high. Carlton is appropriately cautious in his interpretation, noting that energy prices may be acting as proxies for other inputs left out of the model. Carlton's list of variables for each metropolitan area is not overlong; it includes wages, man-hours, unemployment, number of engineers, income taxes, property taxes, and constants related to size of firm and economies of scale—in addition to the prices of electricity and natural gas. Subtler influences on location, such as unionism, the quality of the labor force, and the size or composition of the market, could well be acting but not captured by these variables. In Carlton's defense, however, it should be noted that the estimated relative importance of energy accorded well with actual relative use of energy in the three industries studied.

In the other study the issue of energy and industry location is approached more directly but in a less sophisticated way. Studying newly opened plants of Fortune 500 companies, I asked the relevant plant managers if "low energy costs" were "perceived as 'desirable, if available,' and helped to tip the scales in favor of this site."[11] The question was designed, in a crude way, to elicit the importance of energy prices to

10. Dennis W. Carlton, "Why New Firms Locate Where They Do: An Economic Model," in William Wheaton, ed., *Interregional Movements and Regional Growth* (Urban Institute, 1979); and Carlton, "The Location and Employment Choices of New Firms: An Econometric Model with Discrete and Continuous Endogenous Variables" (University of Chicago, 1980).

11. Schmenner, *Making Business Location Decisions*. This book combines the most important aspects of the two following reports by the same author: "The Location Decisions of Large, Multiplant Companies," report to the U.S. Department of Housing and Urban Development, September 1980; and "The Manufacturing Location Decision: Evidence from Cincinnati and New England," report to the U.S. Department of Commerce, Economic Development Administration, March 1978.

choice of location when all other factors are equal or nearly so. Of the plants from which surveys were returned only 25 percent indicated that low energy costs were "desirable, if available," and significant for the location of the site chosen. That frequency of response was only tenth in a list of thirteen location factors given, suggesting that energy costs are not an important location consideration for all industry. When the responses are disaggregated by industry group, however, low energy costs are found to be more significant for new plant openings in the heavy metals industry (frequency 75 percent) and specialty chemicals or metals (57 percent). The only other industry group in which their influence is above the average is that made up of heavy chemicals, oil, rubber, and glass (29 percent). The order of the other industry groups make sense as well, market-sensitive industries and those tied to forests being the least influenced by low energy costs.

Other work of mine, on companies and plants in the Cincinnati metropolitan area and in New England, also has to do with the effects of energy on the location of industry.[12]

Companies that had opened new plants in New England during the first portion of the 1970s were queried about the reasons for opening new plants rather than expanding on site elsewhere. "Took advantage of better energy availability" was one of nine possible responses. It was by far the least popular response, cited by only two of forty-two companies. Of course, its low showing could be attributed to New England itself rather than to the unimportance of energy. Weakening this explanation somewhat, however, is another finding, this one in response to a question about the reasons for expanding a plant on site rather than opening a new one elsewhere. "Improved energy availability elsewhere not a strong enough consideration" was one of nine possible responses. It, too, was tied for last place among the responses, cited by only one of fifty-four responding companies. Thus, while New England is often considered a region of high-cost energy, the cost is apparently not so high for the collection of industry there that plants expanding on site during the first five years of the 1970s would pay much attention to energy considerations.

Plants that had relocated either within the Cincinnati metropolitan area or within New England were asked whether "energy availability and rates" had improved, grown worse, or remained the same strictly as a result of their move. Not surprisingly, for about two thirds of the plants

12. Ibid.

that had relocated, the availability of energy and the rates charged for it remained the same. Of the remaining plants in the two locations, however, the split was about even between those for which the availability of energy and the rates charged for it improved as a result of their moves and those for which they grew worse. These results reinforce the view that energy considerations are not critical to most plant-location decisions, especially within metropolitan areas.

These two studies notwithstanding, the effect of the availability of process energy or the prices charged for it on the location of industry is still not known with much certainty, and much work remains to be done. Researchers are still not very far removed from the simple, although powerful, observation of industries, such as aluminum, that are known to be heavily dependent on energy. We do not know, for example, for which industries or types of plant energy costs have only recently become enough of a concern to influence the decisions about location of new facilities significantly.

And we can only speculate as to the effects of something such as decontrol of natural gas prices on the location of industry. While we know that manufacturers are more concerned with the availability of natural gas than with its price—and should thus be pleased that decontrol of prices will do much to stabilize supplies—there is likely to be at least some substitution of other sources of energy for natural gas when prices are allowed to rise. Whether this price rise will stimulate any changes of location, other things being equal, is still an open question. Surely manufacturers will be able to pass along much, if not all, of the rise in natural gas prices to their customers, and this will argue against any change of location. But whether any increase in natural gas prices left unshifted will be enough to stimulate a change of location to an area in which natural gas prices or the prices of alternative factors of production are lower is not clear. While I seriously doubt that decontrol of natural gas prices will have any effect on the location of industry, I cannot support this view with any conclusive evidence.

The Effects of Energy Prices and Availability on the Location of Industry: Transportation

Transportation expenses, whether incurred in assembling raw materials or in distributing output, have been widely recognized as important influences on the plant locations of many industries. Industries such as dairy and bakery products, beverages, paperboard boxes, printing, plastics, and paper and metal containers are usually located close to any concentration of population, in part so that costs of transport will be lower and delivery times improved. In such industries transport costs could be considered controlling concerns; some companies—Anheuser-Busch and General Foods, for example—are so sensitive to them that they have employed sophisticated techniques such as linear programming to help assess possible plant locations. In the absence of other controlling concerns, many other industries often prefer locations that put transport costs at a minimum. Similarly, many of the established techniques of plant-location consultants involve a systematic appraisal of transport costs incurred. It is no wonder that as advances have been made in process technologies and communications, as well as in transportation, manufacturing employment of all types has become more heterogeneous geographically. The relative growth of formerly nonindustrial regions such as the South and Southwest and of formerly little industrialized suburbs and rural areas bears out the universal appeal of lower transport costs and the pervasiveness of a host of technological advances.

One summary measure of the geographic dispersion of industry has been devised.[13] Known as the coefficient of localization, it sums the deviations between the percentage distributions of manufacturing value added and personal income across areas. If manufacturing value added is distributed exactly as personal income, the coefficient is 0. If all manufacturing is located in a single area, the coefficient is 1. Calculated coefficients of localization confirm the market sensitivity of such important industries as stone, clay, and glass; paper; food; and printing.

My research on the Fortune 500 provides some insights into the importance of transportation to industries and regions.[14] From table 10, in which market and supply areas are shown, it is clear that not all companies and industries value proximity to markets or supplies. And, as is clear from simple geography and the sorting out of industries across the country, not all regions value proximity in the same way. In general, though, supplies are collected from a closer range than that within which outputs are marketed, and the variation among industries in the supply-

13. See Miller, *Manufacturing*, pp. 110–14.
14. Schmenner, *Making Business Location Decisions*.

Table 10. *Market Areas and Supply Areas, by Industry Group and Region*
Average number of miles shown in all responses to survey

Industry group and region[a]	*Radius within which half of market lies*	*Radius within which three quarters of market lies*	*Radius within which half of supplies lie*	*Radius within which three quarters of supplies lie*
Industry group				
Agriculture-tied	599	1,018	415	596
Market-sensitive	501	747	625	956
Forest-tied	438	656	438	703
Labor-rate sensitive	724	1,230	763	1,138
Heavy chemicals, oil, rubber, and glass	687	1,232	534	966
Specialty chemicals and metals	671	1,238	435	834
Heavy metals	743	978	550	1,097
Industrial machinery and transportation equipment	851	1,520	520	998
High technology	1,587	2,954	800	1,434
Region				
New England	1,254	2,074	589	1,214
Mid Atlantic	754	1,430	452	970
South Atlantic	916	1,360	659	1,052
East North Central	539	1,174	320	659
East South Central	510	822	540	726
West North Central	671	1,000	477	995
West South Central	832	1,505	515	966
Mountain	640	931	950	1,266
Pacific	1,120	2,032	1,066	1,587

Source: Roger W. Schmenner, "The Location Decisions of Large, Multiplant Companies," report to the Department of Housing and Urban Development, September 1980.
a. Industries are grouped according to the standard industrial classification.

area mileage is less than the variation in market-area mileage. These findings, coupled with statements made in interviews, suggest that for most companies local procurement of supplies is more successful than local marketing of outputs. Indeed, as indicated by the survey of industry associations conducted for this paper, the industries that spend more in distributing outputs than in gathering inputs outnumber the industries that do the opposite.

A list of the industries most sensitive to proximity to their markets—food processing, printing, container manufacture, and conversion of forest products, for example—contains no surprises. These industries were most likely to answer that their plants were "seriously impeded from competing in some geographic markets solely because of transport costs." Forty-seven percent of forest-tied plants and 40 percent of the so-called market-sensitive plants gave that answer. In contrast, only 3 percent of the high-technology plants and only 9 percent of the industrial machinery and transportation equipment plants were "seriously impeded" by transport costs. These industries were much more likely to state that transport costs were "only slightly important."

Naturally, the region in which it is located has a bearing on the market and supply areas of a plant. The center of the country offers some obvious geographic access, which is reflected in table 10. It is also true that more plants that are affected by transport costs are to be found in the center of the United States than on its coasts. While for 57 percent of the plants in New England and 45 percent of the plants on the Pacific Coast transport costs are "only slightly important," for plants in the East South Central region, the percentage drops to 29. Differences among other regions, however, are not so sharp.

No matter what the industry or region, the plants for which transport costs are a consideration have smaller market areas than others. The following table shows just how constrained market areas are for some plants.[15]

Importance of cost of transport to plant	*Radius within which half of market lies (miles)*	*Radius within which three quarters of market lies (miles)*
Serious impediment	447	814
Modest impediment	634	1,132
Only slightly important	1,198	1,992

Given the significant increases in the real costs of transport during the past decade, an important question is whether these increases have actually altered the patterns of industrial location. Have companies tried to avoid higher transport costs by locating closer to their markets, perhaps

15. Ibid.

trading higher labor costs for lower freight bills? The question is difficult to answer on several counts:

As noted earlier, technological advances of all kinds have freed manufacturing plants from many former location constraints, and the drift seems to be toward location near markets, although less often within central cities.

Transportation has long been a factor in plant location, and if transport costs were an important consideration in the selection of the present location, increasing those costs is not likely to induce much of a rearrangement.

For at least some industries, higher energy costs make larger plants more attractive, because of the possibilities for cogeneration, economies in heat consumption, or having their own power plants. This trend, however, is in conflict with the trend associated with transport costs toward smaller, market-oriented plants. This conflict may imply a locational standoff for many industries.

To my knowledge, there has been no research on this point. I can only offer some loose, almost anecdotal evidence culled from my own survey of industry trade associations. In that survey I inquired whether the value of being close to either the market or suppliers had changed for companies in the industry since the days before the energy crisis. Of nineteen industries that responded, seven stated that being closer to the market was now more important to them. None stated that it was less important; most acknowledged no change. As could be expected, fewer industries—only four—stated that being closer to suppliers was now more important; the remainder acknowledged no change.

I also asked about the cost of commuting to work—whether it mattered at all in the selection of plant location and if so whether city or suburban sites were now more attractive. For thirteen of the eighteen industries that responded, commuting costs were of no consequence; of the remaining five, four indicated that suburban sites were now more attractive.

Most discussion of manufacturing plants and transportation centers on the costs of either assembling raw materials or distributing output. There are two other purposes of transportation that can influence plant location, namely, moving work-in-process inventory between plants and moving troubleshooting managers and engineers from headquarters to plants. As suggested earlier, the cost and time associated with those activities can frequently act as a controlling concern in the plant-location choice.

If considerable work-in-process inventory is moved from plant to plant or if troubleshooting by management is prevalent, it is likely that the plants will be located in a cluster so as to diminish the costs of transport and handling and improve delivery and commuting times. Such plant clusters are common in shoes, apparel, textiles, defense contracting, and high-technology industries.

Concluding Comments

There is no doubt that the so-called energy crisis of the 1970s had a profound effect on American industry. The response of industry has been to substitute other factors of production for energy, where possible, conserving as much as it reasonably can, and switching from expensive or unreliable forms of energy to less expensive forms or those forms supplies of which are better assured. The speed and extent of industry's response has been remarkable—new fuel-efficient and often more flexible equipment has been designed, constructed, and installed, and new procedures and production processes are being followed.

In part because of this technological success, it is difficult for me to envision significant changes in patterns of industrial location that may have been triggered, or are likely to be triggered, by continued high energy costs. A number of other considerations reinforce this view. First, to most manufacturing companies for which energy is now important, energy has always been important, and thus even their long-standing locations were selected with energy in mind. A site that was energy-effective some years ago, moreover, is still likely to be energy-effective now, suggesting that even a significant increase in energy prices is not likely to prompt many changes of location among such manufacturers. It should be remembered in this fast-paced technological age that only recently—say, within the last fifty years or so—has much of industry been as free as it is to locate without primary regard for energy. It was not that long ago that steel mills were largely confined to locations near coalfields and aluminum production near hydroelectric power. Whereas the locations of traditionally energy-sensitive industries are unlikely to change significantly, the question remains whether industries traditionally not energy-sensitive are now locating plants with energy costs in mind. Answers to this question are difficult and no one has done thorough research on it. In the survey sent to industry trade associations as background for writing this paper, only five of the nineteen that responded seemed to think that prices and supplies of process energy were shifting production in favor of one region or another—to the South and Southwest, for example. In addition, only four of the nineteen seemed to think that transport considerations were shifting production as well—either to the Midwest or the South, for example. These findings are suggestive, but more careful research is clearly required.

Second, although energy price differences still exist, they are narrowing steadily. As the tables introduced earlier show, the variations among states in the distribution of energy prices of all kinds has been declining. This finding suggests that in the future energy prices will be even less of a factor in plant-location decisions than they are now. One caveat persists: what of the availability of energy, distinct from its price? The interruptions of oil and natural gas supplies to which some industries in certain locations have been subjected have elevated the availability of energy as an industrial concern of some consequence. As noted earlier, however, concern about availability has stimulated shifts away from oil and natural gas toward sources of energy considered more reliable, such as coal and electricity. This same concern has fostered an increase in fuel-flexible designs for boilers and other equipment. Choice of location as a way of assuring energy supplies is apparently not held to be effective. What data and informal conversation with manufacturers reveal is that availability may occasionally dictate location on main natural gas or oil pipelines, rather than on smaller branch pipelines, but such considerations are primarily restricted to the largest users, such as petrochemical and fertilizer plants that use natural gas or oil as feedstocks.

Third, the effect of energy considerations on plant location is attenuated by the sometimes conflicting forces that conservation of energy may set in motion. As was noted earlier, process energy efficiency may well imply that fewer, larger plants are to be preferred, while at the same time transport energy efficiency may imply that many small plants located close to the principal markets are desirable. Depending on the industry, of course, one of these forms may override the other, but in at least some industries a standoff is likely.

Fourth, as can be inferred from the data, commuting costs rarely make any difference in the choice of plant location, but if they have any influence today, it is likely to be in favor of suburban rather than urban sites. Central cities have been losing manufacturing employment to suburbs and rural areas for many years and for many reasons—greater needs for space, new production technologies that demand new layouts, the availability of labor, avoidance of congestion and crime, better access to transportation, and lower site-related costs—and energy concerns are unlikely to stem this tide.

Comment by Dennis W. Carlton

Roger Schmenner has provided a useful and thorough investigation of what is known about the effects of energy on location. I agree with Schmenner's main point that for most industries, the recent changes in relative energy prices among regions are unlikely to have dramatic direct effects on patterns of industrial location. I shall address only two points: those factors that determine whether regional discrepancies in energy prices will influence industrial location and the timing of adjustments to increases in regional discrepancies in energy prices.

Does Energy Matter?

Schmenner's paper shows that energy cost divided by value added is quite small for all but a handful of industries with two-digit standard industrial classifications (SIC). Let me propose another descriptive measure—energy cost divided by value of shipments. Consider a firm deciding whether to locate one of its plants in region A or region B. That firm will consider in which region profits are higher. Profits equal revenue minus costs. Observable costs in any region depend on wage rates, energy prices, and the prices of other factors of production. Unobservable factors that influence profits, such as the way a particular location fits into the firm's distribution network, will also influence the choice of location and will vary among firms and regions. It is then possible to write:

$$\pi_A = R_A - C(W_A, EP_A, OP_A) + E_A,$$

and

$$\pi_B = R_B - C(W_B, EP_B, OP_B) + E_B,$$

where subscripts indicate region and where

π = profits,
R = revenue,
C = observable cost,
W = wages,
EP = energy price,
OP = other prices, and
E = idiosyncratic factors.

The question is, If energy prices in region A change, will the preference of the firm for locating in region A instead of region B be altered? That is, will π_A still exceed π_B? Clearly, the more important energy is as a component of total cost, the more likely it is that any given change in energy prices will alter the locational preferences of a firm. This reasoning suggests that it might be worth while to look at energy costs divided by total costs. Using value of shipments as a proxy for total costs, I have calculated energy cost divided by value of shipments for the twenty two-digit SIC-code manufacturing industries (table 11).

As table 11 indicates, the importance of energy as a fraction of total costs is quite small, even in the most energy-intensive industries. Much the same story is to be found in table 12, where the calculations are repeated for those three-digit SIC-code industries identified in Schmenner's paper as having the highest ratio of energy cost to value added. As table 12 indicates, for only a handful of industries is energy cost equal to more than 10 percent of value of shipments, and for none is it equal to more than 25 percent. These small percentages suggest that even large regional disparities in energy prices may have little effect on relative profits.

Of course, such descriptive measures as energy cost divided by value added or by value of shipments can only suggest the effect of energy prices on choice of location. The correct approach is to ask what the likelihood is that a given change in regional energy prices will cause a firm to change its locational preference from region A to region B. Another way to pose this question is to ask how inframarginal firms are in relation to energy prices. Let me explain this point more carefully. A firm must decide whether to locate a plant in region A or region B. In general, not only energy costs but also wages, other costs, revenues, and unobservable effects will differ between the two regions. For region A, consider how profits will change as energy costs increase, holding

Table 11. *Energy Intensity of Selected Large Industries: Energy Costs as a Percent of the Value of Shipments, 1977*

SIC code	*Industry*	*Energy costs (millions of dollars)*	*Value of shipments (millions of dollars)*	*Energy costs as percent of value of shipments*	*Rank*
20	Food processing	2,537.8	192,911.6	1.3	10
21	Tobacco	63.9	9,050.6	0.7	20
22	Textiles	1,136.6	40,550.5	2.8	5
23	Apparel	285.8	40,245.1	0.7	19
24	Lumber	726.5	39,919.4	1.8	8
25	Furniture	198.3	16,978.0	1.2	11
26	Paper	2,960.4	52,085.7	5.7	3
27	Printing	421.7	49,716.2	0.8	18
28	Chemicals	6,448.7	118,153.6	5.4	4
29	Petroleum	2,445.4	97,452.7	2.5	6
30	Rubber and plastics	962.4	39,552.8	2.4	7
31	Leather	82.6	7,607.4	1.1	12
32	Stone, clay, glass, and cement	2,587.4	35,476.6	7.3	1
33	Primary metals	7,043.6	103,179.4	6.8	2
34	Fabricated metals	1,348.7	90,023.5	1.5	9
35	Machinery	1,268.0	122,187.7	1.0	15
36	Electrical and electronic machinery	986.7	88,433.1	1.1	13
37	Transportation equipment	1,420.6	166,954.0	0.9	17
38	Instruments	272.5	28,897.8	0.9	16
39	Miscellaneous manufacturing	199.9	19,150.7	1.0	14

Sources: Bureau of the Census, *Industry Groups and Industries*, p. 4-14; and Bureau of the Census, *Census of Manufactures, 1977: General Summary*, pp. 30–44.

revenues, wages, and other factors unchanged. The response is the downward-sloping relation depicted in figure 1.

The line labeled π_A corresponds to profits in region *A* and lies above the line labeled π_B, which corresponds to profits for region *B*. The line labeled π_A lies above that labeled π_B, because when energy prices are equal, region *A* is more advantageous to the firm than region *B*. If energy prices in the two regions are not equal, the firm decides where to locate by comparing profits at the prevailing price of energy in region *A* with

Table 12. *Energy Costs as a Percent of the Value of Shipments of Selected Three-Digit SIC-Code Industries, 1977*[a]

SIC code	*Industry*	*Energy costs as percent of value of shipments*	*Rank*
207	Fats and oils	1.9	24
214[b]	Tobacco stemming and redrying	1.0	25
226	Dyeing and finishing textiles	5.0	18
249	Miscellaneous wood products (particle board)	4.6	20
261[b]	Pulp mills	9.4	7
262[b]	Paper mills	10.3	6
263[b]	Paperboard	13.9	4
266[b]	Building paper (including asphalt, tar, and asbestos)	15.5	2
281	Industrial inorganic chemicals (chlorine and industrial gases)	14.6	3
282	Plastics materials and synthetic resins and rubber	5.1	17
286[b]	Industrial organic chemicals	7.4	13
287	Agricultural chemicals (fertilizers)	6.4	15
291[b]	Petroleum refining	2.5	23
295	Paving and roofing materials	4.5	21
303[b]	Reclaimed rubber	7.9	12
321[b]	Flat glass	8.6	11
322	Glass and glassware (including bottles)	9.1	9
324[b]	Cement	24.1	1
325	Structural clay products (bricks, refractories)	12.2	5
329	Abrasives and asbestos	4.8	19
331	Blast furnaces and steelworks	8.9	10
332	Iron and steel foundries	6.9	14
333	Primary smelting and refining of nonferrous metals (aluminum)	9.3	8
334[b]	Secondary smelting and refining of nonferrous metals	2.5	22
339	Miscellaneous wood products (particle board)	5.7	16

Sources: Bureau of the Census, *Census of Manufactures, 1977: General Summary* pp. 30–44; and Bureau of the Census, *Industry Groups and Industries*, pp. 4-14–4-30.

a. The industries selected are the twenty-five most energy-intensive industries, measured by energy costs as a percent of value added, as reported in Schmenner's paper.

b. Figures are reported for four-digit SICs only.

those in region *B*. Suppose that this comparison is between points *A* and *B* in figure 1. Since *A* is higher than *B*, the firm will locate in region *A*.

If energy prices in region *A* increase, will the preference for region *A* persist? As drawn, energy prices in region *A* would have to increase to *EP** before a switch in locational preferences would occur. The critical

Figure 1. *Effect on Profits of Increases in the Price of Energy in Two Hypothetical Regions*

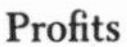

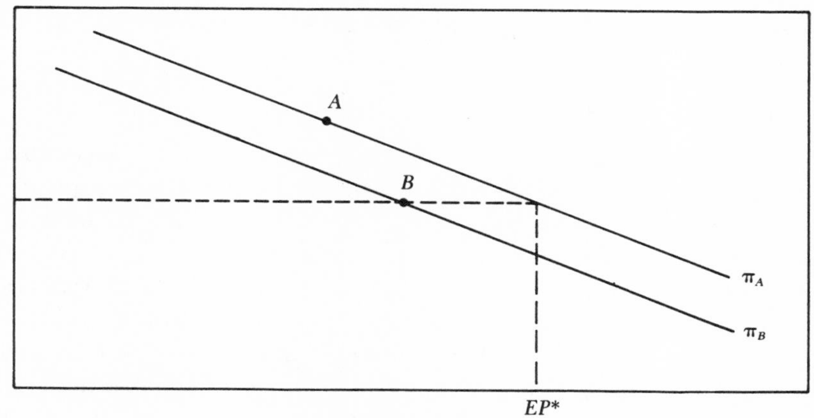

Price of energy

factor determining the effect of changes in energy costs on location is the degree to which a firm in region *A* is inframarginal—that is, how large the gap is between the two profit lines in figure 1. The more inframarginal firms are, the smaller the effect of changing energy prices on choice of location.

From a look at the simple model of locational choice, factors can be identified that tend to make firms more or less inframarginal in relation to changes in energy prices. The greater the variability in wages, in other costs, and in idiosyncratic factors, the greater the tendency of firms to be inframarginal with respect to energy prices and the less the importance of energy as a fraction of profits will be. Wages do vary regionally, and since labor costs form the bulk of the total costs of any industry, wage differences are likely to swamp differences in the cost of energy. My own work and that of Schmenner emphasize the variability of idiosyncratic factors among regions. I have shown that in general energy tends to be only a small component of cost. Schmenner points out, moreover, that regional differences in energy prices have actually diminished in the 1970s. These facts suggest that it will be the exception rather than the rule for changing regional differences in the cost of energy to have significant direct effects on choice of location.

One further point to consider is that many of the most energy-intensive three-digit SIC-code industries seem to be heavy polluters. I wonder whether regions in which energy costs are low would want—or allow—such firms to locate new plants in their areas. Perhaps those industries most likely to move in response to changes in energy costs are—or will be—constrained from doing so by strict zoning ordinances.

Time Pattern of Adjustment

It is complicated to predict how an industry sensitive to energy costs will adjust to higher energy costs. Clearly, if location of a new plant is being considered, the choice of location for that new plant will be based on the changed energy prices. But employment changes very little as a result of a new plant location. It changes most as a result of expansion and contraction of existing plants. The response of existing plants to higher energy costs is not obvious. Firms with existing plants have sunken costs that will prevent them from finding it profitable to move. They may eventually leave a region, but it may take them a long time. In the meantime, moreover, investment or employment may actually increase as optimal adaptation to higher energy costs is made. The problem is even more complicated if local agglomeration effects are introduced. All this suggests how difficult it is to predict the speed with which firms will adapt to changes in energy prices and to use data from the 1970s on *total* changes in employment—rather than new employment owing to new location—to discern the effects of higher energy prices on location. It is possible that the sizable locational shifts in response to differing regional energy price increases are still to come, though on the basis of my analysis presented here, I tend to doubt it.

Conclusion

To say that energy prices appear to have little direct effect on choice of location does not mean that they may not have enormous indirect effects. As energy costs rise, for example, other costs, such as transport costs, may also rise. By influencing the desirability of being close to buyers and suppliers, rising transport costs may exert an important effect on locational decisions. Other indirect, or general equilibrium, effects of energy on wages, local prices, and so on could similarly have significant effects on location decisions. Miernyk's paper suggests that one effect of changing regional energy costs is to change regional income. These income effects could change the demand for products and labor and could influence prices, wage rates, and so on, which would in turn influence choice of location. The purpose of partial equilibrium analysis is to figure out how energy costs will affect choice of location. My discussion and Schmenner's paper indicate that the direct, or partial equilibrium, effects of energy costs on choice of location are likely to be small. The effect, if any, of energy on choice of location is therefore more likely to come through indirect, or general equilibrium, effects of energy on other factors, such as income, which in turn influence location.

WILLIAM H. MIERNYK

Energy and Regional Development

REGIONAL shifts in population and economic activity are as old as the nation. They have been going on since the early postcolonial era. While the rate of change has been far from constant, the direction of movement has held steady. The center of population gravity has moved from east to west. Regional shifts in economic activity are less clear-cut than those of population. But the general direction has been from the northeast to the south and west.

There are important differences in the effects of interregional shifts in a growing national economy and those observed in one that is nearly stagnant. Some areas have failed to benefit, in an economic sense, even during periods of robust national growth. Those areas range in size from multistate regions to counties or even smaller governmental units. They were adversely affected by shifts in demand, factor substitution, technological change, resource depletion, and other economic changes that differ from region to region. Such problems are easier to attack under conditions of robust national growth than when the economy is growing slowly, or perhaps not at all.

Three important regional development programs—the Area Redevelopment Administration, the Economic Development Administration, and the Appalachian Regional Commission—were established during the first half of the 1960s. That was a period of rapid national growth and stable

The author gratefully acknowledges the assistance of Dee Knifong, Lucinda Robinson, Alan Mierke, Jo Alice Evans, James Cassell, Tom Beam, Churai Tapvong, Jean Gallaher, and Carla Uphold. The figures were prepared by Jean Stansberry. Drafts of the paper were greatly improved by the incisive comments of Anthony Downs, Katharine L. Bradbury, Frank Hopkins, and a number of participants in the conference.

prices. Senators and congressmen from prosperous states were willing to support legislation designed to stimulate economic development in bypassed areas or redevelopment in areas that were the victims of industrial decline. Conditions have changed radically since that period, however, and a basic hypothesis of this paper is that the changes are permanent rather than transitory. This is likely to lead to future changes in regional development policy.

The most important recent work on national economic growth is that of Edward F. Denison, who has been called by Richard Stone—himself a pioneer in national income accounting—"the originator of growth accounting."[1] Denison found that between 1929 and 1948 national income grew at an annual rate of 2.49 percent. Between 1948 and 1973 the rate increased to 3.65 percent, but between 1973 and 1976 it dropped to 0.58 percent. A far more important measure of economic growth is what Denison calls "national income per person employed (NIPPE)." The corresponding figures for the three periods mentioned above are 1.21 percent, 2.12 percent, and −0.22 percent.[2] The latter figures are particularly significant because they reflect the influence of changes in productivity on economic growth. They also have much to do with the changing character of the regional problems of the country.

Interregional shifts in population and economic activity do not arise from any single cause. There is an extensive literature on human migration.[3] Models of growing complexity have been developed, but even the most sophisticated cannot include or measure all the "push" and "pull" forces that induce people to relocate. It is even more difficult to isolate the factors responsible for industrial relocation, with a few notable exceptions, among them the massive migration of labor-intensive industries, such as cotton and woolen textiles and full-fashioned hosiery from the north to the south following World War II. Because of the public focus on a variety of energy issues since 1973, however, there has been a tendency to select differential energy prices and availability as the principal causes of recent economic and demographic shifts.

1. Richard Stone, "Whittling Away at the Residual: Some Thoughts on Denison's Growth Accounting," *Journal of Economic Literature*, vol. 18 (December 1980), p. 1539.

2. Edward F. Denison, *Accounting for Slower Economic Growth: The United States in the 1970s* (Brookings Institution, 1979), pp. 104–08.

3. For an excellent summary, see Harry W. Richardson, *Regional Growth Theory* (London: Macmillan, 1973), pp. 89–103; see also the series of articles on migration in *International Regional Science Review*, vol. 6 (Fall 1981).

There has been a regional "OPEC," or income-transfer, effect. It is less pronounced at the interregional level than it has been at the international level, but it is still significant. Interregional effects are far more difficult to measure with any degree of accuracy than international effects. Accurate records of international transactions are kept, but the absence of data on interregional trade is one of the better-known discontents of regional economists and regional scientists. Even if highly accurate data on interregional trade were available, moreover, it would be difficult to disentangle the causal strands of shifts in economic activity because of their interdependence.

In spite of these caveats, there is no doubt that differential changes in energy prices have exacerbated the problems of older industrial areas and have stepped up the tempo of regional shifts. The rapid rise of energy prices during the 1970s has led to expressions of concern about their negative effects on older industrial areas and their potential consequences for the future. A report by the Advisory Commission on Intergovernmental Relations stated that:

> the Commission is keenly aware of the growing concern for the economic growth prospects of energy "have," contrasted with "have not," states. It recognizes that as the energy rich states gain revenues from exportable severance taxes on gas, oil, and coal they may replace income, sales, and property taxes paid by resident individuals and businesses. If this happens, state-local tax differentials could become far more significant for industrial location than they now appear to be. The Commission intends to monitor this development actively to detect any new source of divergence in regional and state economic growth.[4]

An analysis of the potential effects on energy-surplus states of severance taxes, royalty payments, and other income effects of rising energy prices has been made by Corrigan and Stanfield.[5] They, too, feel that interstate differentials in both prices and the availability of energy will have an effect on the future location of industry: "For the Frostbelt states, a major worry is that their lack of energy resources will accelerate the shift of industry and jobs to the energy-rich regions."[6] Corrigan and Stanfield recognize that other factors influence the location of industry, but they conclude that "while energy has been only one factor . . . its importance

4. Advisory Commission on Intergovernmental Relations, *Regional Growth: Interstate Tax Competition,* ACIR A-76 (Washington, D.C.: ACIR, 1981), p. iii.

5. Richard Corrigan and Rochelle L. Stanfield, "Rising Energy Prices: What's Good for Some States Is Bad for Others," *National Journal,* vol. 12 (March 22, 1980), pp. 468–74.

6. Ibid., p. 474.

undoubtedly has grown."[7] Elsewhere Corrigan has estimated that the oil-exporting states will realize windfall gains of more than $127 billion during the next decade.[8] These estimates no doubt reinforce the fears expressed by such organizations as the Advisory Commission on Intergovernmental Relations. More will be said about severance taxes and their policy implications in a later section.

Concerns about the potential effects of rising energy costs on older industrial areas are understandable, but location decisions are rarely based on a single element of spatially variable cost.[9] The relative importance of such costs differs from sector to sector. In theory, a location decision is made by selecting a site at which the algebraic sum of expected spatially variable costs is at a minimum.

It is unusual to find a case in which a single element of spatially variable cost is dominant—that is, where the positive cost differential of one input is greater than the sum of negative differentials of all other elements of spatially variable cost. When assertions are made that factories are moving to low-cost energy sites, there is an implicit assumption that the amount saved on energy will more than offset any additions to operating and transfer costs incurred by the relocation.[10] It is far more likely, when actual relocations take place, that there are several factors in the attraction of a new location and that no single element of cost is dominant.

None of the foregoing is meant to imply that the fears expressed about negative effects of rising energy prices on older industrial areas are groundless. In my view, however, the primary effects are obscured by focusing on industrial location or relocation. The central thesis of this paper is that rising energy prices have altered the terms of trade between energy-surplus and energy-deficit states and that this shift has had more to do with interregional income transfers during the past decade than the relocation of manufacturing industries. Before this thesis can be argued, however, a number of preliminary matters must be discussed. The first is the regional classification used in this paper.

7. Ibid.

8. Richard Corrigan, "Economic War over Oil Also Looms at the State Level," *National Journal*, vol. 11 (December 22, 1979), p. 2138.

9. Relatively few costs are independent of location. Some exceptions are federal taxes, certain insurance premiums, and similar expenses that are uniform throughout the country.

10. In the jargon of location theory, the term "transfer costs" refers to the costs of shipping raw materials and fuels to a specific site plus the cost of shipping products to customers. Transfer costs should not be confused with relocation costs.

Figure 1. *The Standard Federal Regions*

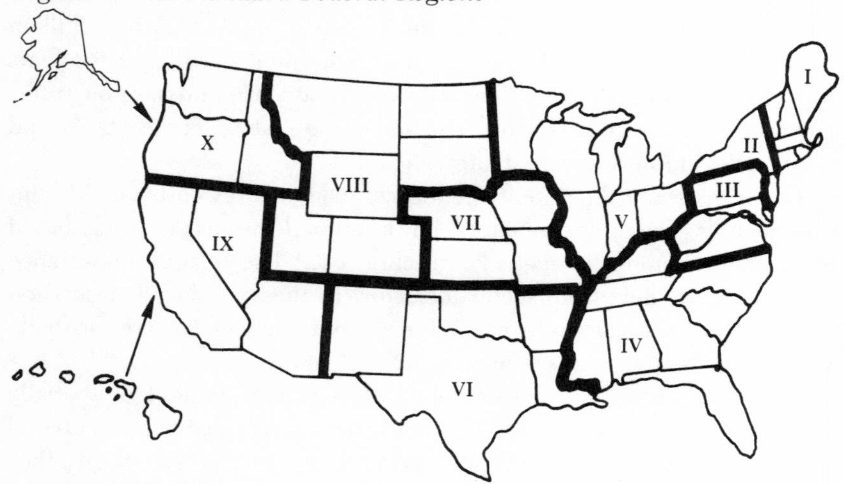

Region I: New England
Connecticut, Maine, Massachusetts, New Hampshire, Rhode Island, Vermont

Region II: North Mid-Atlantic
New Jersey, New York

Region III: South Mid-Atlantic
Delaware, District of Columbia, Maryland, Pennsylvania, Virginia, West Virginia

Region IV: Southeast
Alabama, Florida, Georgia, Kentucky, Mississippi, North Carolina, South Carolina, Tennessee

Region V: Great Lakes
Illinois, Indiana, Michigan, Minnesota, Ohio, Wisconsin

Region VI: Southwest
Arkansas, Louisiana, New Mexico, Oklahoma, Texas

Region VII: Central Plains
Iowa, Kansas, Missouri, Nebraska

Region VIII: Rocky Mountain–Northern Plains
Colorado, Montana, North Dakota, South Dakota, Utah, Wyoming

Region IX: Far West
Arizona, California, Hawaii, Nevada

Region X: Northwest
Alaska, Idaho, Oregon, Washington

Source: General Services Administration, Office of the Federal Register, *United States Government Manual, 1982/83* (GPO, 1982), app. D, p. 845.

The Choice of Regional Boundaries

No division of a country so large and diversified as the United States into regions will be satisfactory for all analytical purposes. It would be convenient, for example, to define regions by degree of energy self-sufficiency, but the result would be sets of noncontiguous states and would be more confusing than illuminating. Whatever the choice, there is some degree of arbitrariness in any delineation of regions. The one chosen for this paper is that used to establish the standard federal regions (SFRs).[11] These regions and their constituent states are given in figure 1.

No SFR is made up entirely of energy-surplus states, although the Southwest and the Rocky Mountain–Northern Plains regions come close. The former includes Arkansas, a relatively small state with a modest energy deficit. The only deficit state in the latter is South Dakota, again a small state with a small deficit. On the other hand, some of the SFRs consist entirely of energy-deficit states—that is, states that produce less energy than they consume.

The basic geographic cut in this paper is the one that distinguishes between energy-surplus and energy-deficit areas, and that is best made at the state level. In the interests of brevity and consistency, however, the discussion of energy self-sufficiency and its economic consequences is conducted at the regional level. (The energy-surplus or deficit status of each state is given in table 11.)

Regional Shifts in Population, Employment, and Income

Relative changes in population, by state, between 1970 and 1980, are given in table 1. The long-run drift from the Northeast to the South and West continued during this decade. There were absolute declines in New York, Rhode Island, and the District of Columbia while increases

11. The SFRs were established by executive order in February 1972. A Federal Regional Council, made up of the principal regional officials of major federal departments, was set up in each of the ten regions. The objective was "to achieve more uniformity" in the location and geographic jurisdiction of federal field offices; see Office of the Federal Register, National Archives and Record Service, *United States Government Manual, 1978–79* (Government Printing Office, 1980), app. D, p. 842.

Table 1. *Population Change, by Region and State, 1970 and 1980*

Region and state	*Population* 1970	*Population* 1980	*Percent change, 1970–80*	*Percent of U.S. total* 1970	*Percent of U.S. total* 1980
New England	11,847,245	12,348,493	4.2	5.8	5.5
Connecticut	3,032,217	3,107,576	2.5	1.5	1.4
Maine	993,722	1,124,660	13.2	0.5	0.5
Massachusetts	5,689,170	5,737,037	0.8	2.8	2.5
New Hampshire	737,681	920,610	24.8	0.4	0.4
Rhode Island	949,723	947,154	−0.3	0.5	0.4
Vermont	444,732	511,456	15.0	0.2	0.2
North Mid-Atlantic	25,412,503	24,921,446	−1.9	12.5	11.0
New Jersey	7,171,112	7,364,158	2.7	3.5	3.3
New York	18,241,391	17,557,288	−3.8	9.0	7.8
South Mid-Atlantic	23,425,120	24,611,973	5.1	11.5	10.9
Delaware	548,104	595,225	8.6	0.3	0.3
District of Columbia	756,668	637,651	−15.7	0.4	0.3
Maryland	3,923,897	4,216,446	7.5	1.9	1.9
Pennsylvania	11,800,766	11,866,728	0.6	5.8	5.2
Virginia	4,651,448	5,346,279	14.9	2.3	2.4
West Virginia	1,744,237	1,949,644	11.8	0.9	0.9
Southeast	31,862,549	38,860,776	22.0	15.7	17.2
Alabama	3,444,354	3,890,061	12.9	1.7	1.7
Florida	6,791,418	9,739,992	43.4	3.3	4.3
Georgia	4,587,930	5,464,265	19.1	2.3	2.4
Kentucky	3,220,711	3,661,433	13.7	1.6	1.6
Mississippi	2,216,994	2,520,638	13.7	1.1	1.1
North Carolina	5,084,411	5,874,429	15.5	2.5	2.6
South Carolina	2,590,713	3,119,208	20.4	1.3	1.4
Tennessee	3,926,018	4,590,750	16.9	1.9	2.0
Great Lakes	44,068,850	45,746,886	3.8	21.7	20.2
Illinois	11,110,285	11,418,461	2.8	5.5	5.0
Indiana	5,195,392	5,490,179	5.7	2.6	2.4
Michigan	8,881,826	9,258,344	4.2	4.4	4.1
Minnesota	3,806,103	4,077,148	7.1	1.9	1.8
Ohio	10,657,423	10,797,419	1.3	5.2	4.8
Wisconsin	4,417,821	4,705,335	6.5	2.2	2.1
Southwest	20,343,132	25,043,102	23.1	10.0	11.1
Arizona	1,923,322	2,285,513	18.8	0.9	1.0
Louisiana	3,644,637	4,203,972	15.3	1.8	1.9
New Mexico	1,017,055	1,299,968	27.8	0.5	0.6
Oklahoma	2,559,463	3,025,266	18.2	1.3	1.3
Texas	11,198,655	14,228,383	27.1	5.5	6.3
Central Plains	11,237,395	11,764,045	4.7	5.5	5.2
Iowa	2,825,368	2,913,387	3.1	1.4	1.3
Kansas	2,249,071	2,363,208	5.1	1.1	1.0
Missouri	4,677,623	4,917,444	5.1	2.3	2.2
Nebraska	1,485,333	1,570,006	5.7	0.7	0.7

Table 1 *(continued)*

Region and state	Population 1970	Population 1980	Percent change, 1970–80	Percent of U.S. total 1970	Percent of U.S. total 1980
Rocky Mountain–Northern Plains	5,579,743	6,950,250	24.6	2.7	3.1
Colorado	2,209,596	2,888,834	30.7	1.1	1.3
Montana	694,409	786,690	13.3	0.3	0.3
North Dakota	617,792	652,695	5.6	0.3	0.3
South Dakota	666,257	690,178	3.6	0.3	0.3
Utah	1,059,273	1,461,037	37.9	0.5	0.6
Wyoming	332,416	470,816	41.6	0.2	0.2
Far West	23,005,119	28,150,612	22.4	11.3	12.4
Arizona	1,775,399	2,717,866	53.1	0.9	1.2
California	19,971,069	23,668,562	18.5	9.8	10.4
Hawaii	769,913	965,000	25.3	0.4	0.4
Nevada	488,738	799,184	63.5	0.2	0.4
Northwest	6,520,375	8,107,242	24.3	3.2	3.6
Alaska	302,583	400,481	32.4	0.1	0.2
Idaho	713,015	943,935	32.4	0.4	0.4
Oregon	2,091,533	2,632,663	25.9	1.0	1.2
Washington	3,413,244	4,130,163	21.0	1.7	1.8
All United States	203,302,031	226,504,825	11.4	100.0	100.0

Source: U.S. Bureau of the Census, *Census of Population and Housing, 1980*, Advance Report PHC80-V-1 (Government Printing Office, 1980), table 1.

in population in Ohio, Pennsylvania, and Massachusetts were less than 2 percent. Florida was the only eastern state to register a gain of 35 percent or more, although New Hampshire and South Carolina were in the group of states that gained between 20 and 35 percent. Other states with similar population gains were in the Southwest, the Rocky Mountain–Northern Plains, and the Northwest. The last was the only region in which all constituent states registered population gains of at least 20 percent.

Total employment, by SFR, and changes in employment between 1940 and 1979—with one intervening year—are given in table 2. Employment has grown in all regions since 1940, but growth has varied widely among regions. Throughout the period, growth rates were below average in New England, the Great Lakes, the Middle Atlantic regions, and the Central Plains. The pattern was essentially the same between 1970 and 1979, although the Central Plains region barely moved out of the below-average category. The Far West grew the most rapidly during the entire period, but this region slipped into fourth place after 1970,

Table 2. *Employment, by Standard Federal Region, 1940, 1970, and 1979*

Region	Employment			Percent change		Percent of U.S. total		
	1940	1970	1979	1940–79	1970–79	1940	1970	1979
New England	3,060,127	4,799,974	5,765,000	88.4	20.1	6.7	6.2	5.9
North Mid-Atlantic	6,543,577	10,053,077	10,730,000	64.0	6.7	14.4	12.9	11.1
South Mid-Atlantic	5,784,757	8,959,721	10,507,000	81.6	17.3	12.7	11.5	10.8
Southeast	7,071,088	12,367,822	15,401,000	117.8	24.5	15.6	15.9	15.9
Great Lakes	10,188,311	17,115,057	20,449,000	100.7	19.5	22.5	21.9	21.1
Southwest	4,292,449	7,267,117	10,179,000	137.1	40.1	9.5	9.3	10.5
Central Plains	3,177,128	4,341,003	5,465,000	72.0	25.9	7.0	5.6	5.6
Rocky Mountain–Northern Plains	1,175,654	2,048,539	3,064,000	160.6	49.6	2.6	2.6	3.2
Far West	2,897,712	8,664,227	11,998,000	314.1	38.5	6.4	11.1	12.4
Northwest	1,185,012	2,391,745	3,460,000	192.0	44.7	2.6	3.1	3.6
All United States	45,375,815	78,008,282	96,945,000	113.6	24.3	100.0	100.0	100.0

Sources: For 1940, 48 contiguous states: Bureau of the Census, *Census of the Population*, vol. 3, *The Labor Force*, pts. 2–5 (GPO, 1943), table 17; Alaska and Hawaii: Bureau of Economic Analysis, *Regional Employment by Industry, 1940–1970* (GPO, 1975), table 8; 1970, Bureau of the Census, *Census of Population, 1970*, vol. 1, *Characteristics of the Population* (GPO, 1973), table 183; 1979, Bureau of the Census, *Statistical Abstract of the United States, 1980* (GPO, 1980), table 656. Figures for 1979 are rounded to the nearest thousand: detail for 1979 does not add to U.S. total because data for states are derived from independent population controls.

following the Rocky Mountain–Northern Plains, the Northwest, and the Southwest, in that order.

Shares of total employment are given in the three right-hand columns. The largest declines were in the Mid-Atlantic regions, while the Far West almost doubled its share between 1940 and 1979. After 1970, there were continued declines in New England and the Middle Atlantic and Great Lakes regions, stability in the Southeast, and increases in the remaining regions. It is not surprising that, in general, shifts in total employment roughly parallel the shifts in population described in table 1.

As has employment, per capita income has increased in every region. Table 3 shows changes between 1929 and 1980, with one intermediate year. Data for 1980 are presented both in current dollars and in dollars adjusted for regional differences in the cost of living (COL). Adjustment factors were calculated from 1977 estimates of "real" state income per capita made by the National Center for Economic Alternatives (NCER).[12] Reliable data on state-to-state differences in the cost of living are not available, so the adjusted 1980 data must be regarded as estimates. It is not likely, however, that they are subject to a wide margin of error.

The right-hand side of table 3 shows a strong tendency toward convergence in unadjusted income per capita among regions between 1929 and 1980. But only one region—the South Mid-Atlantic—moved from one side of the U.S. average to the other, and that change was from a fraction above to a fraction below the average.

The 1980 estimates of real per capita income suggest some reswitching of relative positions among regions. The two northeastern regions dropped below the U.S. average. Other regions in which cost-of-living adjustments produced a decline in real per capita income are the Great Lakes, the Far West, and the Northwest. Only in the last of these was the change large enough to cause a switch from above-average to below-average status.

The South Mid-Atlantic region, with below-average unadjusted per

12. News release, "Living Cost Differences Shift Income Ranking of States," National Center for Economic Alternatives, January 7, 1979. "Real" income in the SFRs was estimated by taking percentage differences between actual 1977 state income per capita and the adjusted estimates made by the NCER. These were applied to 1980 data to obtain adjusted 1980 income per capita by state. The latter were multiplied by state population to obtain state personal income. Data on state population and income were aggregated to the SFR level to obtain the estimates given in table 3.

Table 3. *Per Capita Income, by Region, 1929, 1959, and 1980*

Region	Per capita income (dollars)				Percent of U.S. total			
			1980				1980	
	1929	1959	Current dollars	Adjusted for differences in the cost of living	1929	1959	Current dollars	Adjusted for differences in the cost of living
New England	876	2,339	10,146	9,030	124.6	108.7	106.6	96.3
North Mid-Atlantic	1,103	2,654	10,456	9,306	156.9	123.4	109.8	99.3
South Mid-Atlantic	709	2,111	9,495	9,623	100.8	98.1	99.7	102.7
Southeast	352	1,562	7,961	8,632	50.1	72.6	83.6	92.1
Great Lakes	784	2,296	9,774	9,560	111.5	106.7	102.7	102.0
Southwest	435	1,801	9,014	9,781	61.9	83.7	94.7	104.4
Central Plains	590	2,039	9,327	9,399	83.9	94.8	98.0	100.3
Rocky Mountain–Northern Plains	532	1,941	9,076	9,254	75.7	90.2	95.3	98.7
Far West	964	2,585	10,696	10,139	137.1	120.2	112.3	108.2
Northwest	691	2,232	9,847	9,301	98.3	103.8	103.4	99.2
All United States	703	2,151	9,521	9,373	100.0	100.0	100.0	100.0

Sources: For 1929, U.S. Department of Commerce, Office of Business Economics, *Personal Income by States since 1929*, Supplement to the *Survey of Current Business* (GPO, 1956), pp. 140–45 (data for Hawaii and Alaska are not included); for 1959, U.S. Department of Commerce, Bureau of Economic Analysis, *Survey of Current Business*, vol. 46 (August 1966), pp. 12, 13; Bureau of the Census, *Statistical Abstract of the United States, 1967* (GPO, 1967), p. 12; and for 1980, per capita income, *Survey of Current Business*, vol. 61 (July 1981), p. 31; population, Bureau of the Census, *Census of Population and Housing, 1980*, Advance Report PHC80-V-1, table 1; cost-of-living adjustment factor calculated from news release, "Living Cost Differences Shift Income Ranking of States," National Center for Economic Alternatives, January 7, 1979.

capita income, moved above the average after the cost-of-living adjustment. This was a result of rising coal prices and below-average living costs in the region except in Pennsylvania. That combination also explains the improvement in real per capita income in the Central Plains and the Rocky Mountain–Northern Plains. The substantial improvement in the Southeast, however, is entirely the result of differences in the cost of living.

The most dramatic switch was that of the Southwest. In unadjusted dollars, per capita income in this region rose from less than 70 percent of the U.S. average in 1929 to almost 95 percent in 1980. After the cost-of-living adjustment, however, it moved up to 104 percent of the U.S. average. Lower living costs contributed to the upward shift in real income. The dominant factor, however, was the rise of oil and gas prices in this energy-rich region.

The Production and Consumption of Energy, by Region

To compare production and consumption of energy by region different forms of basic energy must be converted to a common unit, typically the British thermal unit. Aggregate national data are reported in "quads," or quadrillion Btu. Regional data, however, are more accurately expressed in trillions of Btu, the basic measure used in this paper.[13] The production and consumption of basic energy in the United States—by type, SFR, and economic sector—are given in a series of tables with a common format.

Energy Production, by Region

The production of energy, by type and SFR, is summarized in table 4. Coal and natural gas are the most important domestically produced sources of energy, each accounting for almost a third of the total. Petroleum ranks third. Hydroelectric energy stands in fourth place among the sources of energy that can be allocated by region. The sources of nuclear fuel and imported fossil fuels, which together account for 20

13. Energy specialists do not agree completely on conversion factors. Those used to make estimates given in this paper were published in American Petroleum Institute, *Basic Petroleum Data Book: Petroleum Industry Statistics* (Washington, D.C.: API, 1981), section 25, table 2.

Table 4. *Production of Energy, by Region, 1979*
Trillions of Btu

Region	Coal[a]	Natural gas[b]	Petroleum[c]	Hydro-electric energy[d]	Geo-thermal energy[d]	Energy from wood and waste[d]	Total[e]	Percent of U.S. total
New England	0.000	0.000	0.000	80.327	0.000	0.115	80.442	0.12
Percent of row	0.00	0.00	0.00	99.86	0.00	0.14	100.00	...
Percent of column	0.00	0.00	0.00	2.53	0.00	3.29	0.13	...
North Mid-Atlantic	0.000	16.042	4.959	324.866	0.000	0.000	345.867	0.53
Percent of row	0.00	4.64	1.43	93.93	0.00	0.00	100.00	...
Percent of column	0.00	0.08	*	10.24	0.00	0.00	0.56	...
South Mid-Atlantic	6,416.118	256.016	30.646	49.294	0.000	0.000	6,752.074	10.38
Percent of row	95.02	3.79	0.45	0.73	0.00	0.00	100.00	...
Percent of column	31.48	1.26	0.17	1.55	0.00	0.00	10.88	...
Southeast	4,701.328	325.691	636.745	343.265	0.000	0.000	6,007.029	9.23
Percent of row	78.26	5.42	10.60	5.71	0.00	0.00	100.00	...
Percent of column	23.07	1.60	3.52	10.82	0.00	0.00	9.68	...
Great Lakes	3,421.982	264.814	425.272	135.327	0.000	0.213	4,247.608	6.53
Percent of row	80.56	6.23	10.01	3.19	0.00	*	100.00	...
Percent of column	16.79	1.30	2.35	4.26	0.00	6.09	6.84	...
Southwest	1,257.862	17,505.966	10,149.656	51.838	0.000	0.997	28,966.319	44.53
Percent of row	4.34	60.44	35.04	0.18	0.00	*	100.00	...
Percent of column	6.17	86.02	56.06	1.63	0.00	28.52	46.66	...

Central Plains	206.980	781.621	366.292	32.753	0.000	0.090	1,387.736	2.13
Percent of row	14.91	56.32	26.39	2.36	0.00	*	100.00	...
Percent of column	1.02	3.84	2.02	1.03	0.00	2.57	2.24	...
Rocky Mountain–Northern Plains	3,924.760	727.757	1,471.220	281.432	0.000	0.540	6,405.709	9.85
Percent of row	61.27	11.36	22.97	4.39	0.00	*	100.00	...
Percent of column	19.26	3.58	8.13	8.87	0.00	15.45	10.32	...
Far West	298.418	246.060	2,053.054	479.520	64.350	0.170	3,141.572	4.83
Percent of row	9.50	7.83	65.35	15.26	2.05	*	100.00	...
Percent of column	1.46	1.21	11.34	15.11	100.00	4.86	5.06	...
Northwest	153.532	227.753	2,965.743	1,394.376	0.000	1.371	4,742.775	7.29
Percent of row	3.24	4.80	62.53	29.40	0.00	*	100.00	...
Percent of column	0.76	1.12	16.38	43.95	0.00	39.22	7.64	...
Nuclear[f]	...	...	...	...	...	...	2,976.585	4.58
Total U.S. production	20,380.980	20,351.720	18,103.587	3,172.998	64.350	3.496	65,053.716	100.00
Percent of row	31.33	31.28	27.83	4.88	0.10	*	100.00	...
Percent of column	100.00	100.00	100.00	100.00	100.00	100.00	100.00	...
Imported fuels[f]	...	...	...	...	...	...	13,071.840	...
Total U.S. consumption	...	...	...	...	...	...	78,125.554	...

Sources: Coal, U.S. Department of Energy, Energy Information Administration, *Energy Data Report: Coal Production, 1979* (GPO, 1981), table 1, p. 5; natural gas, DOE, EIA, *Energy Data Report: Natural Gas Production and Consumption, 1979* (GPO, 1981), table 4, p. 8; petroleum, DOE, EIA, *Energy Data Report: Crude Petroleum, Petroleum Products, and Natural Gas Liquids* (GPO, 1980), tables 3 and 4, p. 8; and hydroelectric energy, geothermal energy, and energy from wood and waste, DOE, EIA, *State Energy Data Report* (GPO, April 1980), p. 3.

* Less than 0.05 percent.

a. Mines producing less than 10,000 short tons during the year were excluded. The conversion factor for coal is 1 short ton = 26,200,000 Btu.

b. Natural gas data are for dry natural gas—marketed production less extraction loss. The conversion factor is 1,000 cubic feet = 1,035,000 Btu.

c. The conversion factor for petroleum is 1 barrel = 5,800,000 Btu.

d. Production of hydroelectric energy, geothermal energy, and energy from wood and waste is assumed to equal consumption at the source.

e. Percentages do not include nuclear fuels.

f. Could not be allocated by region of origin or destination.

percent of total consumption, cannot be identified. (An estimate of the total amount of energy from the last-named sources is given in the lower right corner of table 4 to bring national production and consumption into balance.)

New England and the North Mid-Atlantic region produce negligible quantities of energy. They are clearly the most dependent upon imports—from other regions and other countries—of all the SFRs. The largest exporter is the Southwest, which accounted for more than 46 percent of national energy production in 1979. Four of the five states in the region are energy exporters, and the fifth, Arkansas, is relatively small.

Four regions—the South Mid-Atlantic, the Southeast, the Great Lakes, and the Rocky Mountain–Northern Plains—account for 91 percent of national coal production. The Southwest alone produces 86 percent of the natural gas, the Rocky Mountain region and the Northern and Central Plains together producing an additional 7.4 percent. The remainder is spread more or less evenly over the other SFRs, excluding New England and the North Mid-Atlantic states.

The Southwest also produces more than half the nation's oil—56 percent. Other important oil-producing areas are the Northwest, which includes Alaska, the Rocky Mountain–Northern Plains, and the Far West. Smaller amounts are produced in the Southeast and the Great Lakes.

Every region produces some hydroelectric energy, but the Northwest, with 44 percent, is the largest producer. The Far West, with 15 percent, is in second place, followed by the North Mid-Atlantic and the Southeast, each of which produces more than 10 percent of the U.S. total. This total, however, is only 5 percent of national energy production. All the nation's geothermal energy was produced in the Far West in 1979, but geothermal energy accounted for only 2 percent of the energy produced in that region. Wood and waste provided some energy in all but three regions in 1979; the amounts, however, were quite small.

Energy Consumption, by Region

The United States consumed more than 78 quadrillion Btu, or quads, of energy in 1978. Almost two thirds of this total was consumed in four regions—the Great Lakes, the South Mid-Atlantic, the Southeast, and the Southwest. The shares used by other regions are given in table 5, which also shows consumption, by principal category, in each region.

Electrical utilities are the largest consumer of basic energy, accounting for 30 percent of national consumption. The range across regions is surprisingly wide, however, going from less than 25 percent in the Southwest to more than 47 percent in the Northwest. Utilities are unique in that they convert basic energy to a form used directly by consumers, in industrial processes, and to provide commercial and transportation services. About 65 percent of the energy consumed by utilities is "used up" in the process of generating and transmitting electricity. The end-use distribution of energy is substantially altered if the energy consumed by utilities is "passed through" to end-use sectors. Table 6 shows the results of this adjustment.

Industry, which accounts for almost 37 percent of nonelectric end use, is the dominant user of energy in the distribution given in table 6. The range of industrial use varies from 21 percent of the regional totals in New England and the North Mid-Atlantic states, to 52 percent in the Southwest. The only other region to depart from the U.S. average by more than a few percentage points is the Far West, where industry accounted for about 26 percent of the end use of energy.[14]

Households used about 21 percent of all energy consumed in the United States in 1978, while commercial establishments used 16 percent. The two northeastern regions used considerably more than these averages, reflecting, in part, the greater use of energy for heating. Residential use in the Great Lakes and Central Plains regions, which include a number of states along the northern tier, was also above average. The lowest residential share—13 percent—was in the Southwest. Relatively little fuel is needed for home heating in this region.

The contrast between the Great Lakes and the Southwest is noteworthy. Both are large industrial consumers of energy, each accounting for almost a fourth of the national total. The Great Lakes region also used a fourth of national household energy, but the Southwest, because of lower heating requirements, used only 11 percent.

Except for New England and the North Mid-Atlantic states, in which use is above the national average, there is not much regional variation

14. The unusually large share of energy consumed by industry in the Southwest is explained in part by the concentration of gas utilities and petroleum refineries—the two most energy-intensive industrial sectors in the country—in that region. Directly and indirectly they require almost 53 cents worth of inputs per dollar of output. The manufacture of paving mixtures and blocks ranks third in energy intensity with a total energy input coefficient of 23 cents per dollar of output. See William H. Miernyk, Frank Giarratani, and Charles F. Socher, *Regional Impacts of Rising Energy Prices* (Ballinger, 1978), p. 82.

Table 5. *Consumption of Energy by the Total End-Use Sector, by Region, 1978*[a]
Trillions of Btu

Region	*Residential*	*Commercial*[b]	*Industrial*[c]	*Transportation*	*Electric utilities*	*Total energy consumed*[c]	*Percent of U.S. total*
New England	565.164	498.640	429.881	844.278	865.589	3,203.552	4.10
Percent of row	17.64	15.57	13.42	26.35	27.02	100.00	. . .
Percent of column	6.68	7.85	2.23	4.10	3.68	4.10	. . .
North Mid-Atlantic	1,116.364	901.790	782.695	1,644.830	1,583.943	6,029.622	7.72
Percent of row	18.51	14.96	12.98	27.28	26.27	100.00	. . .
Percent of column	13.20	14.20	4.05	7.99	6.73	7.72	. . .
South Mid-Atlantic	870.786	612.187	2,069.557	1,947.362	2,829.958	8,329.850	10.66
Percent of row	10.45	7.35	24.85	23.38	33.97	100.00	. . .
Percent of column	10.30	9.64	10.71	9.46	12.02	10.66	. . .
Southeast	803.697	699.997	2,412.903	3,609.091	4,740.024	12,265.712	15.70
Percent of row	6.55	5.71	19.67	29.42	38.64	100.00	. . .
Percent of column	9.50	11.03	12.49	17.53	20.13	15.70	. . .
Great Lakes	2,578.917	1,588.350	4,298.790	3,687.271	4,652.060	16,805.388	21.51
Percent of row	15.35	9.45	25.58	21.94	27.68	100.00	. . .
Percent of column	30.50	25.02	22.26	17.91	19.76	21.51	. . .

Southwest	762.504	649.182	5,621.962	3,156.326	3,305.782	13,495.756	17.27
Percent of row	5.65	4.81	41.66	23.39	24.49	100.00	. . .
Percent of column	9.02	10.23	29.11	15.33	14.04	17.27	. . .
Central Plains	642.767	446.458	1,034.311	1,183.702	1,150.142	4,457.380	5.71
Percent of row	14.42	10.02	23.20	26.56	25.80	100.00	. . .
Percent of column	7.60	7.03	5.35	5.75	4.89	5.71	. . .
Rocky Mountain–Northern Plains	320.779	285.296	732.100	729.400	991.023	3,058.598	3.91
Percent of row	10.49	9.33	23.94	23.85	32.40	100.00	. . .
Percent of column	3.79	4.49	3.79	3.54	4.21	3.91	. . .
Far West	652.826	464.855	1,266.066	2,927.261	1,864.173	7,175.181	9.18
Percent of row	9.10	6.48	17.65	40.80	25.98	100.00	. . .
Percent of column	7.72	7.32	6.55	14.21	7.92	9.18	. . .
Northwest	142.693	201.931	536.814	863.316	1,559.751	3,304.505	4.23
Percent of row	4.32	6.11	16.24	26.13	47.20	100.00	. . .
Percent of column	1.69	3.18	2.78	4.19	6.63	4.23	. . .
All United States	8,456.496	6,348.688	19,315.833	20,592.841	23,542.447	78,125.554	100.00
Percent of row	10.82	8.13	24.72	26.36	30.13	100.00	. . .
Percent of column	100.00	100.00	100.00	100.00	100.00	100.00	. . .

Source: Department of Energy, Energy Information Administration, *State Energy Data Report*, p. 4.

a. Without electricity distributed.

b. Includes municipal government and institutions estimated to account for about 0.7 percent of national consumption.

c. Does not include net imports of 130.754 trillion Btu of coal coke that cannot be allocated to the states.

Table 6. *Consumption of Energy by the Nonelectric End-Use Sector, by Region, 1978*
Trillions of Btu

Region	*Residential*	*Commercial*[a]	*Industrial*[b]	*Transportation*	*Total energy consumed*[b]	*Percent of U.S. total*
New England	895.915	792.648	687.950	845.075	3,221.588	4.12
Percent of row	27.81	24.60	21.35	26.23	100.00	...
Percent of column	5.48	6.31	2.40	4.10	4.12	...
North Mid-Atlantic	1,593.614	1,491.138	1,291.387	1,648.311	6,024.450	7.71
Percent of row	26.45	24.75	21.44	27.36	100.00	...
Percent of column	9.76	11.87	4.51	7.99	7.71	...
South Mid-Atlantic	1,752.697	1,317.301	3,129.228	1,951.443	8,150.670	10.43
Percent of row	21.50	16.16	38.39	23.94	100.00	...
Percent of column	10.73	10.49	10.94	9.46	10.43	...
Southeast	2,627.085	1,739.146	4,323.467	3,612.939	12,302.635	15.75
Percent of row	21.35	14.14	35.14	29.37	100.00	...
Percent of column	16.08	13.85	15.11	17.52	15.75	...
Great Lakes	4,044.668	2,692.634	6,634.918	3,693.222	17,065.443	21.84
Percent of row	23.70	15.78	38.88	21.64	100.00	...
Percent of column	24.76	21.44	23.19	17.91	21.84	...

Southwest	1,810.764	1,452.614	6,972.838	3,161.274	13,397.490	17.15
Percent of row	13.52	10.84	52.05	23.60	100.00	...
Percent of column	11.09	11.57	24.37	15.33	17.15	...
Central Plains	1,075.326	780.390	1,391.632	1,184.784	4,432.132	5.67
Percent of row	24.26	17.61	31.40	26.73	100.00	...
Percent of column	6.58	6.21	4.86	5.74	5.67	...
Rocky Mountain–Northern Plains	527.046	452.856	1,003.176	730.302	2,713.380	3.47
Percent of row	19.42	16.69	36.97	26.91	100.00	...
Percent of column	3.27	3.61	3.51	3.54	3.47	...
Far West	1,389.599	1,328.444	2,029.140	2,931.923	7,679.106	9.83
Percent of row	18.10	17.30	26.42	38.18	100.00	...
Percent of column	8.51	10.58	7.09	14.22	9.83	...
Northwest	618.079	511.379	1,144.588	864.596	3,138.640	4.02
Percent of row	19.69	16.29	36.47	27.55	100.00	...
Percent of column	3.78	4.07	4.00	4.19	4.02	...
All United States	16,334.792	12,558.550	28,608.324	20,623.873	78,125.540	100.00
Percent of row	20.91	16.07	36.62	26.40	100.00	...
Percent of column	100.00	100.00	100.00	100.00	100.00	...

Source: Department of Energy, Energy Information Administration, *State Energy Data Report*, p. 4.

a. Includes municipal government and institutions estimated to account for about 0.7 percent of national consumption.

b. Does not include imports of 130.754 trillion Btu of coal coke that cannot be allocated to the states.

in the consumption of energy by commercial establishments. That in the Southwest is lower than the average, because such a large share of its energy is consumed by its industrial sector.

Transportation used 26 percent of the total energy consumed in the United States in 1978. The only region that differed significantly from the national average in this share was the Far West, where transportation accounted for 38 percent of regional energy consumption.

The sources of residential energy are shown in table 7. Natural gas was the primary source in 1978, accounting for 30 percent of the national total. The share used in the Northwest was equal to only a third of the national average, however, while the shares used in New England and the Southeast were equal to half that average. The largest users of natural gas were the Great Lakes region and the Far West.

Oil contributed 21 percent to residential use in 1978. Considerably more than this average was used in the northeastern regions, with shares of 47 and 40 percent. The largest deviation in the other direction was in the Far West, where petroleum contributed less than 5 percent to household energy use. Use in the Southwest and the Northwest was also well below average, the former because of low household heating requirements, the latter doubtless because of the availability of relatively inexpensive hydroelectricity.

Electric utilities provided about 14 percent of all the energy consumed by U.S. households in 1978. In doing so, they used 34 percent of total household energy. The largest deviations above the national average were in the Southeast and the Northwest, in the former because of the availability in the past of steam coal at favorable delivered prices and the influence of TVA and in the latter because of the relative abundance of hydroelectric energy.

Industrial sources of energy are shown in table 8. Natural gas and oil accounted for 55 percent of the industrial fuel used. Coal and electric energy added 12 and 10 percent, respectively, and almost a fourth of all industrial energy consumed was used to generate and transmit electricity.

The largest users of coal were the South Mid-Atlantic and Great Lakes regions, which also accounted for much of U.S. coal production. As might be expected, the Southwest used more natural gas, in both absolute and relative terms, than any other region; in fact, it consumed more than 51 percent of the national total in 1978. Petroleum is the principal industrial fuel in New England and in the North Mid-Atlantic states. Except for those regions and the Southwest, in which use was lower than the U.S.

average, there was not much regional variation in the use of industrial petroleum.

Hydroelectric energy provided less than half of 1 percent of U.S. industrial needs for energy in 1978. New England accounted for 37 percent of all hydroenergy used by industry, but that represents a large share of a relatively small total. Even in New England, hydroelectric energy satisfied less than 2 percent of the industrial requirements.

As noted, electric utilities are enormous consumers of energy. The sources of utility fuel are given in table 9. Coal is far and away the most important boiler fuel in the country, accounting for 44 percent of the total in 1978. Utilities in the South Mid-Atlantic, Southeast, and Great Lakes regions—which produce the bulk of the nation's coal—consumed three fourths of all utility coal in 1978. Other relatively large users of steam coal are the Rocky Mountain–Northern Plains and Central Plains regions. The remaining regions used little steam coal in relation to other boiler fuels.

Petroleum accounted for 16 percent of the fuel consumed by utilities in 1978. In New England and the North Mid-Atlantic states, the petroleum shares were 55 and 46 percent, respectively. Utilities in the South Mid-Atlantic states and the Far West were the only other major consumers of oil.

The Northwest accounted for the largest share—44 percent—of hydroelectric energy used by utilities. Utilities in the North Mid-Atlantic, Southeast, and Far West regions derived 10 percent or more of their energy requirements from hydro sources. The amounts in the remaining regions were smaller.

Nuclear fuels provided almost 13 percent of the energy needs of utilities in 1978. They accounted for more than a third of the fuel required by utilities in New England, however, and this region used 10 percent of the national total. Nuclear fuels contributed a fifth of the energy used by utilities in the North Mid-Atlantic region, or about 11 percent of the national total. The Southeast and Great Lakes states accounted for more than half the nuclear energy consumed by utilities. The South Mid-Atlantic region was also an important consumer of nuclear fuel, using about 17 percent of the national total. In all the last three regions, nuclear fuel accounted for 16 percent or more of the energy consumed by utilities. The western and southwestern regions used little nuclear fuel, whether measured in relative or in absolute terms.

The commercial sector, which uses about 16 percent of the energy

Table 7. *Consumption of Energy by the Residential Sector, by Region, 1978*
Trillions of Btu

Region	Total coal[a]	Natural gas	Petroleum	Electricity: Sales	Electricity: Generated and transmitted	Total energy consumed	Percent of U.S. total
New England	0.249	140.261	424.652	95.853	234.896	895.915	5.48
Percent of row	*	15.66	47.40	10.70	26.22	100.00	. . .
Percent of column	0.23	2.82	12.61	4.20	4.20	5.48	. . .
North Mid-Atlantic	6.409	474.792	635.164	138.310	338.940	1,593.614	9.76
Percent of row	0.40	29.79	39.96	8.68	21.27	100.00	. . .
Percent of column	5.95	9.53	18.86	6.06	6.06	9.76	. . .
South Mid-Atlantic	40.451	492.411	337.923	255.584	626.327	1,752.697	10.73
Percent of row	2.31	28.09	19.28	14.58	35.74	100.00	. . .
Percent of column	37.57	9.88	10.04	11.19	11.19	10.73	. . .
Southeast	11.207	399.145	393.342	528.430	1,294.958	2,627.085	16.08
Percent of row	0.43	15.19	14.97	20.11	49.30	100.00	. . .
Percent of column	10.41	8.01	11.68	23.14	23.14	16.08	. . .
Great Lakes	31.261	1,696.388	851.267	424.784	1,040.968	4,044.668	24.76
Percent of row	0.77	41.94	21.05	10.50	25.74	100.00	. . .
Percent of column	29.03	34.05	25.28	18.60	18.60	24.76	. . .

Southwest	0.072	522.928	239.504	303.792	744.468	1,810.764	11.09
Percent of row	*	28.88	13.23	16.78	41.11	100.00	. . .
Percent of column	0.07	10.50	7.11	13.31	13.31	11.09	. . .
Central Plains	4.828	393.694	244.245	125.358	307.200	1,075.326	6.58
Percent of row	0.45	36.61	22.71	11.66	28.57	100.00	. . .
Percent of column	4.48	7.90	7.25	5.49	5.49	6.58	. . .
Rocky Mountain–Northern Plains	9.261	207.240	104.279	59.776	146.489	527.046	3.23
Percent of row	1.76	39.32	19.79	11.34	27.80	100.00	. . .
Percent of column	8.60	4.16	3.10	2.62	2.62	3.23	. . .
Far West	0.072	588.574	64.180	213.522	523.253	1,389.599	8.51
Percent of row	*	42.36	4.62	15.37	37.65	100.00	. . .
Percent of column	0.07	11.82	1.91	9.35	9.35	8.51	. . .
Northwest	3.856	66.021	72.816	137.769	337.617	618.079	3.78
Percent of row	0.62	10.68	11.78	22.29	54.62	100.00	. . .
Percent of column	3.58	1.33	2.16	6.03	6.03	3.78	. . .
All United States	107.670	4,981.454	3,367.372	2,283.180	5,595.116	16,334.792	100.00
Percent of row	0.66	30.50	20.61	13.98	34.25	100.00	. . .
Percent of column	100.00	100.00	100.00	100.00	100.00	100.00	. . .

Source: Department of Energy, Energy Information Administration, *State Energy Data Report*, p. 5.

* Less than 0.05 percent.

a. Includes bituminous, lignite, and anthracite.

Table 8. *Consumption of Energy by the Industrial Sector, by Region, 1978*
Trillions of Btu

Region	Total coal[a]	Natural gas	Petroleum	Hydro-electric	Electricity		Total energy consumed	Percent of U.S. total
					Sales	Generated and transmitted		
New England	2.496	51.536	362.632	13.217	74.789	183.279	687.950	2.40
Percent of row	0.36	7.49	52.71	1.92	10.87	26.64	100.00	...
Percent of column	0.07	0.60	5.05	37.17	2.74	2.74	2.40	...
North Mid-Atlantic	138.887	136.495	504.765	2.547	147.422	361.269	1,291.387	4.51
Percent of row	10.75	10.57	39.09	0.20	11.42	27.98	100.00	...
Percent of column	4.05	1.60	7.03	7.16	5.40	5.40	4.51	...
South Mid-Atlantic	976.009	396.955	690.908	5.684	307.099	752.573	3,129.228	10.94
Percent of row	31.19	12.69	22.08	0.18	9.81	24.05	100.00	...
Percent of column	28.48	4.65	9.62	15.99	11.25	11.25	10.94	...
Southeast	500.076	670.969	1,236.733	5.125	553.694	1,356.869	4,323.467	15.11
Percent of row	11.57	15.52	28.61	0.12	12.81	31.38	100.00	...
Percent of column	14.59	7.86	17.22	14.41	20.28	20.27	15.11	...
Great Lakes	1,349.759	1,465.082	1,477.990	5.958	677.027	1,659.105	6,634.918	23.19
Percent of row	20.34	22.08	22.28	0.09	10.20	25.01	100.00	...
Percent of column	39.38	17.16	20.58	16.76	24.79	24.79	23.19	...

Southwest	80.328	4,392.416	1,149.163	0.055	391.491	959.385	6,972.838	24.37
Percent of row	1.15	62.99	16.48	*	5.61	13.76	100.00	. . .
Percent of column	2.34	51.44	16.00	0.15	14.34	14.34	24.37	. . .
Central Plains	99.671	397.331	537.297	0.011	103.553	253.767	1,391.632	4.86
Percent of row	7.16	28.55	38.61	*	7.44	18.24	100.00	. . .
Percent of column	2.91	4.65	7.48	0.03	3.79	3.79	4.86	. . .
Rocky Mountain–Northern Plains	167.443	236.213	328.082	0.362	78.560	192.516	1,003.176	3.51
Percent of row	16.69	23.55	32.70	*	7.83	19.19	100.00	. . .
Percent of column	4.89	2.77	4.57	1.02	2.88	2.88	3.51	. . .
Far West	82.742	552.992	629.424	0.907	221.143	541.931	2,029.140	7.09
Percent of row	4.08	27.25	31.02	*	10.90	26.71	100.00	. . .
Percent of column	2.41	6.48	8.76	2.55	8.10	8.10	7.09	. . .
Northwest	30.183	239.095	265.849	1.688	176.136	431.636	1,114.588	4.00
Percent of row	2.64	20.89	23.23	0.15	15.39	37.71	100.00	. . .
Percent of column	0.88	2.80	3.70	4.75	6.45	6.45	4.00	. . .
All United States	3,427.595	8,539.083	7,182.845	35.556	2,730.917	6,692.330	28,608.326	100.00
Percent of row	11.98	29.85	25.11	0.12	9.55	23.39	100.00	. . .
Percent of column	100.00	100.00	100.00	100.00	100.00	100.00	100.00	. . .

Source: Department of Energy, Energy Information Administration, *State Energy Data Report*, p. 7.
* Less than 0.05 percent.
a. Includes bituminous, lignite, and anthracite.

Table 9. *Consumption of Energy by the Electric Utilities, by Region, 1978*
Trillions of Btu

Region	*Total coal*[a]	*Natural gas*	*Petroleum*	*Hydro-electric energy*	*Nuclear electric energy*	*Geo-thermal energy*	*Energy from wood and waste*	*Total energy consumed*	*Percent of U.S. total*
New England	20.659	1.650	474.219	67.110	301.834	0.000	0.115	865.589	3.68
Percent of row	2.39	0.19	54.79	7.75	34.87	0.00	*	100.00	...
Percent of column	0.20	0.05	12.44	2.14	10.14	0.00	3.26	3.68	...
North Mid-Atlantic	208.760	2.060	729.137	322.320	321.665	0.000	0.000	1,583.943	6.73
Percent of row	13.18	0.13	46.03	20.35	20.30	0.00	0.00	100.00	...
Percent of column	2.04	0.06	19.12	10.27	10.81	0.00	0.00	6.73	...
South Mid-Atlantic	1,709.048	3.575	574.867	43.610	498.857	0.000	0.000	2,829.958	12.02
Percent of row	60.39	0.13	20.31	1.54	17.63	0.00	0.00	100.00	...
Percent of column	16.67	0.11	15.08	1.39	16.76	0.00	0.00	12.02	...
Southeast	2,661.129	228.078	734.172	338.139	778.505	0.000	0.000	4,740.024	20.13
Percent of row	56.14	4.81	15.49	7.13	16.42	0.00	0.00	100.00	...
Percent of column	25.96	6.92	19.25	10.78	26.15	0.00	0.00	20.13	...
Great Lakes	3,336.350	77.585	335.683	129.368	772.831	0.000	0.243	4,652.060	19.76
Percent of row	71.72	1.67	7.22	2.78	16.61	0.00	*	100.00	...
Percent of column	32.55	2.35	8.80	4.12	25.96	0.00	6.89	19.76	...

Southwest	557.423	2,355.229	284.134	51.783	56.215	0.000	0.997	3,305.782	14.04
Percent of row	16.86	71.25	8.60	1.57	1.70	0.00	*	100.00	...
Percent of column	5.44	71.44	7.45	1.65	1.89	0.00	28.27	14.04	...
Central Plains	789.697	172.015	59.387	32.742	96.210	0.000	0.090	1,150.142	4.89
Percent of row	68.66	14.96	5.16	2.85	8.37	0.00	*	100.00	...
Percent of column	7.70	5.22	1.56	1.04	3.23	0.00	2.55	4.89	...
Rocky Mountain– Northern Plains	654.416	39.998	8.440	281.071	6.559	0.000	0.540	991.023	4.21
Percent of row	66.03	4.04	0.85	28.36	0.66	0.00	0.05	100.00	...
Percent of column	6.38	1.21	0.22	8.96	0.22	0.00	15.31	4.21	...
Far West	240.050	391.157	607.351	478.613	82.483	64.350	0.170	1,864.173	7.92
Percent of row	12.88	20.98	32.58	25.67	4.42	3.45	*	100.00	...
Percent of column	2.34	11.87	15.93	15.25	2.77	100.00	4.82	7.92	...
Northwest	73.273	25.359	5.636	1,392.688	61.424	0.000	1.371	1,559.751	6.63
Percent of row	4.70	1.63	0.36	89.29	3.94	0.00	0.09	100.00	...
Percent of column	0.71	0.77	0.15	44.39	2.06	0.00	38.87	6.63	...
All United States	10,250.806	3,296.708	3,813.027	3,137.444	2,976.585	64.350	3.527	23,542.447	100.00
Percent of row	43.54	14.00	16.20	13.33	12.64	0.27	*	100.00	...
Percent of column	100.00	100.00	100.00	100.00	100.00	100.00	100.00	100.00	...

Source: Department of Energy, Energy Information Administration, *State Energy Data Report*, p. 9.
* Less than 0.05 percent.
a. Includes bituminous, lignite, and anthracite.

used in the United States, consumed more petroleum—28 percent—than other forms of energy, followed by natural gas, which satisfied 21 percent of its energy needs (see table 10). A small amount of coal—1.25 percent—was used by this sector. Almost 50 percent of the energy consumed commercially was in the form of electricity, 35 percent being used in generation and transmission.

Use of natural gas varied from 9 percent in New England to 33 percent in the Central Plains. New England was the largest consumer of petroleum, in its commercial sector—54 percent—while the Far West used the smallest amount, 14 percent. There was less variation in end use of electricity, which ranged from about 11 percent in the northeastern regions to almost 19 percent in the Far West.

The transportation sector used about 26 percent of all the energy consumed in the United States in 1978. It will come as no surprise that 97 percent of this was petroleum. Almost all the rest—2.6 percent—was natural gas. In the interest of brevity and because there is relatively little variation among regions, details of transportation energy consumption are not given. The range was from about 93 percent of total use by the sector in the Southwest and Central Plains to 99.8 percent in New England.

Energy-Surplus and Energy-Deficit Regions

The energy-surplus or energy-deficit positions of the Standard Federal Regions are summarized in figure 2, which was derived from tables 4 and 5. The largest deficit area is the Great Lakes region, which includes two important coal-producing states, Indiana and Illinois, and which was responsible for 6.5 percent of national energy production in 1979. It consumed 21.5 percent of the national total in that year, however. The North Mid-Atlantic region, which produces a negligible share of national energy—0.53 percent—ranked second on the deficit side. This region, which includes only New York and New Jersey, consumed 7.7 percent of the national total in 1978.

The Southeast, which includes one energy-surplus state, Kentucky—the number one coal producer in the United States—is the third-largest deficit area. Although this region was responsible for 9 percent of the energy produced in 1979, it consumed 15.7 percent.

The Far West, New England, and Central Plains registered somewhat smaller deficits. The first produced 4.8 percent of the national output in

Figure 2. *Energy-Deficit and Energy-Surplus Regions of the United States*[a]

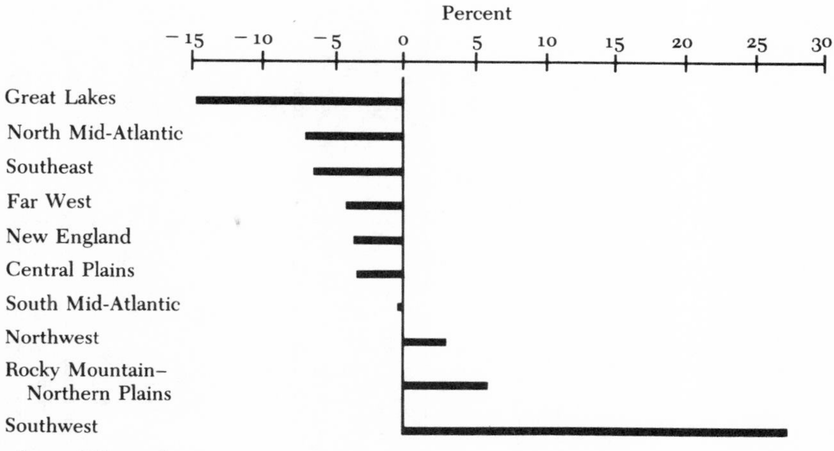

Source: Tables 4 and 5.
a. Percent of U.S. energy produced less percent consumed.

1979 and used 9.2 percent of that consumed nationally. New England produced the smallest share of national output of any of the SFRs—0.12 percent—and consumed 4.1 percent of the national total. The Central Plains produced 2.1 percent and used 5.7 percent of the national totals.

The South Mid-Atlantic region is almost self-sufficient in energy. It includes West Virginia, which ranks second in national coal production, as well as Pennsylvania, Virginia, and Maryland, where significant amounts of coal are also produced. This region contributed 10.3 percent of the national energy output in 1979 and consumed 10.4 percent, to register a slight deficit.

There were moderate surpluses in the Northwest and the Rocky Mountain–Northern Plains. The former accounted for 7.3 percent of national output and 4.2 percent of national consumption. The latter produced 9.9 percent and consumed 3.9 percent of the national totals—the smallest share consumed in any of the SFRs.

The Southwest stands out as the largest surplus area in the country. It includes only one state—Arkansas—which is not an exporter of energy. Louisiana, Oklahoma, Texas, and New Mexico exported substantial quantities of energy in 1979. This region consumed 17.3 percent of the national total, but was responsible for 44.5 percent of the national output.

Four of the SFRs—New England, the North Mid-Atlantic, the Great

Table 10. *Consumption of Energy by the Commercial Sector, by Region, 1978*
Trillions of Btu

Region	Total coal[a]	Natural gas	Petroleum	Electricity: Sales	Electricity: Generated and transmitted	Total energy consumed	Percent of U.S. total
New England	0.168	72.997	425.476	85.205	208.802	792.648	6.31
Percent of row	*	9.21	53.68	10.75	26.34	100.00	...
Percent of column	0.11	2.76	12.00	4.73	4.73	6.31	...
North Mid-Atlantic	6.203	194.049	701.538	170.797	418.551	1,491.138	11.87
Percent of row	0.42	13.01	47.05	11.45	28.07	100.00	...
Percent of column	3.95	7.34	19.77	9.49	9.42	11.87	...
South Mid-Atlantic	40.654	222.515	349.018	204.346	500.768	1,317.301	10.49
Percent of row	3.09	16.89	26.49	15.51	38.01	100.00	...
Percent of column	25.87	8.42	9.84	11.35	11.35	10.49	...
Southeast	20.814	273.531	405.652	301.150	737.995	1,739.146	13.85
Percent of row	1.20	15.73	23.32	17.32	42.43	100.00	...
Percent of column	13.24	10.35	11.43	16.73	16.73	13.85	...
Great Lakes	56.149	809.094	723.107	320.029	784.254	2,692.634	21.44
Percent of row	2.09	30.05	26.86	11.89	29.13	100.00	...
Percent of column	35.73	30.62	20.38	17.78	17.78	21.44	...

Southwest	0.134	345.738	303.311	232.840	570.592	1,452.614	11.57
Percent of row	*	23.80	20.88	16.03	39.28	100.00	...
Percent of column	0.09	13.08	8.55	12.94	12.94	11.57	...
Central Plains	8.968	257.199	180.293	96.776	237.156	780.390	6.21
Percent of row	1.15	32.96	23.10	12.40	30.39	100.00	...
Percent of column	5.71	9.73	5.08	4.38	5.38	6.21	...
Rocky Mountain–Northern Plains	17.201	133.209	134.890	48.560	119.000	452.856	3.61
Percent of row	3.80	29.42	29.79	10.72	26.28	100.00	...
Percent of column	10.94	5.04	3.80	2.70	2.70	3.61	...
Far West	0.134	275.297	189.424	250.273	613.315	1,328.444	10.58
Percent of row	*	20.72	14.26	18.84	46.17	100.00	...
Percent of column	0.09	10.42	5.34	13.91	13.91	10.58	...
Northwest	7.162	59.095	135.673	89.680	219.768	511.379	4.07
Percent of row	1.40	11.56	26.53	17.54	42.98	100.00	...
Percent of column	4.56	2.24	3.82	4.98	4.98	4.07	...
All United States	157.162	2,642.724	3,548.383	1,799.657	4,410.204	12,558.550	100.00
Percent of row	1.25	21.04	28.25	14.33	35.12	100.00	...
Percent of column	100.00	100.00	100.00	100.00	100.00	100.00	...

Source: Department of Energy, Energy Information Administration, *State Energy Data Report*, p. 6.
* Less than 0.05 percent.
a. Includes bituminous, lignite, and anthracite.

Lakes, and the Far West—do not include any energy-surplus states. Each of the others includes at least one. Regional production and consumption have been compared in order to maintain consistency in this paper. The SFRs provide a useful division of energy-deficit and energy-surplus areas, one that is sufficient for examining the basic thesis of this paper. The same information at the state level is more revealing, however, and this is given in table 11.

Regional Variation in Energy Prices

Rising energy prices affect consumers both directly and, when they increase the cost of producing goods and providing services, indirectly. The periods chosen for this and subsequent price comparisons start with a year when energy prices were declining in relation to the general price level. The terminal year was dictated by the availability of data. By 1978, however, all energy prices were rising more rapidly than the general price level.

Rising retail gasoline prices evoke strong negative responses from consumers because of the extensive use of automobiles in this country for business and pleasure. It is interesting, however, that there is less regional variation in retail gasoline prices than in the prices of other forms of energy.

Average prices, in cents per gallon, are given for the United States and the ten SFRs in table 12. The latest available data are already out of date; they do not reflect the retail increases that followed the 1978 OPEC oil price increases. There is no reason to believe, however, that recent price increases would substantially alter the pattern of regional variation shown in table 12, since that pattern was fairly stable between 1960 and 1978.

Retail gasoline prices tended to converge between 1960 and 1978. The range in 1960 was from 93 percent of the U.S. average to 111 percent. By 1978, the range had narrowed to 97–103 percent. The lowest prices, in 1978, were in the four eastern SFRs, while the highest were still in the three western SFRs.

The pattern of utility coal prices, with substantial variation among regions, presents a sharp contrast (see table 13). Also, the spread between lowest and highest regional prices is widening. The lowest delivered price for coal is in the Rocky Mountain–Northern Plains region. Relatively little coal was mined in the West in 1965, but production increased rapidly between that year and 1978. In the latter year the average delivered price in the Rocky Mountain–Northern Plains region was less than $13 a ton, while the national average was almost $20 a ton. It had also dropped from 70 percent to 64 percent of the U.S. average.

Most of the coal produced in the West is surface-mined, and production costs are low compared with those in the predominantly deep-mined coal-producing areas of the Great Lakes region and Appalachia. Relatively low prices in the Southwest are also explained by the extensive use of surface-mining methods.

New England has consistently had the highest delivered price for steam coal, and the North Mid-Atlantic region moved from third place to second between 1965 and 1978. The Southeast, which includes Kentucky, the number one coal-producing state in the country, ranked third in 1978, followed by the South Mid-Atlantic region, which includes West Virginia and Pennsylvania, both important coal-producing states. In 1965, the delivered price of coal in the Southeast was well below the national average—89 percent—but by 1978 it was almost 69 percent higher than the national average. To a lesser extent, a similar switch took place in the Great Lakes region, which includes Illinois and Indiana, both of which are important coal-producing states. Rising delivered coal prices in the Southeast and Great Lakes regions in recent years have stimulated the expansion of nuclear capacity in these areas.

The surge in coal prices is primarily a demand phenomenon. National coal prices, which had been virtually constant for twenty years, increased rapidly in 1968–69, well before world oil prices started to spiral upward. When petroleum prices quadrupled, in 1973–74, coal prices increased even more rapidly.[15]

The demand for steam coal is a derived demand; it depends on the demand for electric energy. Until the early 1970s, U.S. consumption of electric energy grew at an annual rate of about 7 percent; it doubled approximately every decade. Until the price of steam coal started to rise sharply in 1968, the prices of electric energy fell in relation to the general price level. Since the early 1970s the growth rate of the consumption of electricity has dropped below the long-term trend as rapidly rising prices have led to modest conservation efforts.

15. For details, see ibid., p. 10; Miernyk, "Coal," in Gerald G. Somers, ed., *Collective Bargaining: Contemporary American Experience* (Madison, Wis.: Industrial Relations Research Association, 1980), p. 6.

Table 11. *Energy-Deficit and Energy-Surplus States, 1979, and Changes in Relative Income per Capita, 1959–79*

Region and state	Percent of U.S. total energy[a]		Energy production minus consumption	Relative per capita income (U.S. = 100)		Percent change in relative per capita income, 1959–79
	Consumption	Production		1959	1979	
New England	4.12	0.13	−3.99	108.74	101.56	−6.60
Connecticut	1.01	0.01	−1.00	124.71	115.46	−7.42
Maine	0.47	0.08	−0.39	82.37	80.23	−2.60
Massachusetts	1.88	0.00	−1.88	109.81	101.37	−7.69
New Hampshire	0.30	0.02	−0.28	96.44	95.19	−1.30
Rhode Island	0.28	0.00	−0.28	99.68	97.00	−2.69
Vermont	0.18	0.02	−0.16	80.47	83.54	3.82
North Mid-Atlantic	7.71	0.56	−7.15	123.38	106.44	−13.73
New Jersey	2.51	0.00	−2.51	121.89	111.10	−8.85
New York	5.20	0.55	−4.65	123.14	103.77	−15.73
South Mid-Atlantic	10.43	10.88	0.45	98.14	98.92	0.80
Delaware	0.26	0.00	−0.26	125.49	106.31	−15.28
District of Columbia	b	b	b	135.49	120.48	−11.08
Maryland	1.69	0.14	−1.55	104.99	106.36	1.31
Pennsylvania	5.54	4.14	−1.40	101.62	97.55	−4.01
Virginia	1.94	1.59	−0.35	81.91	97.88	19.50
West Virginia	1.01	5.01	4.00	73.30	84.03	14.64
Southeast	15.75	9.68	−6.07	72.38	86.06	18.90
Alabama	2.13	1.47	−0.66	67.79	79.36	17.07
Florida	3.09	0.50	−2.59	89.59	97.41	8.73
Georgia	2.15	0.06	−2.09	74.46	86.97	16.80
Kentucky	1.55	6.38	4.83	71.82	84.24	17.29
Mississippi	1.07	0.59	−0.48	55.67	70.42	26.50
North Carolina	2.16	0.09	−2.07	69.88	84.18	20.46
South Carolina	1.29	0.05	−1.24	62.19	80.44	29.35
Tennessee	2.31	0.52	−1.79	70.89	83.70	18.07
Great Lakes	21.84	6.84	−15.00	106.74	103.65	−2.90
Illinois	5.40	2.70	−2.70	119.44	111.47	−6.67
Indiana	3.31	1.21	−2.10	101.76	97.69	−4.00
Michigan	3.87	0.70	−3.17	104.16	107.18	2.90
Minnesota	1.80	0.15	−1.65	93.48	101.05	8.10
Ohio	5.48	2.04	−3.44	105.32	99.34	−5.68
Wisconsin	1.99	0.04	−1.95	99.58	96.71	−2.88
Southwest	17.15	46.66	29.51	83.59	94.72	13.32
Arkansas	1.20	0.41	−0.79	63.72	79.03	24.03
Louisiana	4.22	16.41	12.19	77.09	86.44	12.13
New Mexico	0.62	3.27	2.65	88.71	86.17	−2.86
Oklahoma	1.67	4.52	2.85	83.53	96.99	16.11
Texas	9.44	22.05	12.61	88.52	100.17	13.16
Central Plains	5.67	2.24	−3.43	94.79	98.43	3.84
Iowa	1.39	0.04	−1.35	90.19	99.99	10.87
Kansas	1.39	1.82	0.43	96.02	105.24	9.60
Missouri	2.10	0.29	−1.81	97.22	94.05	−3.26
Nebraska	0.80	0.08	−0.72	91.44	98.99	8.26

Table 11 *(continued)*

Region and state	*Percent of U.S. total energy*[a] *Con-sumption*	*Pro-duction*	*Energy production minus con-sumption*	*Relative per capita income (U.S. = 100)* *1959*	*1979*	*Percent change in relative per capita income, 1959–79*
Rocky Mountain–Northern Plains	3.47	10.32	6.85	90.24	95.27	5.57
Colorado	1.21	1.40	0.19	101.62	103.98	2.32
Montana	0.49	1.95	1.46	93.01	87.59	−5.83
North Dakota	0.33	1.04	0.71	71.12	93.82	31.92
South Dakota	0.31	0.12	−0.19	67.98	84.98	25.01
Utah	0.66	0.87	0.21	89.13	82.04	−7.96
Wyoming	0.48	4.93	4.45	103.38	113.10	9.40
Far West	9.83	5.06	−4.77	119.90	112.61	−6.08
Arizona	1.00	0.60	−0.40	90.14	96.01	6.51
California	8.13	4.42	−3.71	122.67	114.52	−6.64
Hawaii	0.32	0.00	−0.32	97.73	105.13	7.57
Nevada	0.37	0.04	−0.33	128.04	119.92	−6.34
Northwest	4.02	7.64	3.62	103.77	105.04	1.22
Alaska	0.40	5.19	4.79	116.10	127.88	10.15
Idaho	0.51	0.17	−0.34	86.63	86.30	−0.38
Oregon	1.16	0.54	−0.62	101.39	102.39	0.99
Washington	1.95	1.75	−0.20	107.27	109.03	1.64
All United States	100.00	83.27	−16.73	100.00	100.00	...

Sources: See tables 3, 4, and 5.
a. Data on consumption are for 1978; data for production are for 1979.
b. Included with Maryland.

The regional pattern of the prices of electric energy is affected by regional variations in coal prices, but the two regional patterns are far from identical, since power plants use a variety of fuels in addition to coal. The prices of electric energy are influenced by the regulatory process. Regulation not only contributes to interregional price differences, it also affects price differentials among the three categories of customers. Table 14 shows wide variation in the prices of electric energy among regions and among residential, commercial, and industrial users as well.

The regional spread in the cost of electric energy to households was fairly narrow in 1965. It ranged from 91 percent of the U.S. average in the Southeast to 117 percent in the Far West. But increases in the prices of residential energy were far from uniform between 1965 and 1978. By the latter year the spread had increased from 72 percent of the U.S. average in the Northwest to almost 135 percent in the North Mid-Atlantic states.

Table 12. *Average U.S. Retail Prices of Gasoline, by Region, 1960 and 1978*
Cents per gallon

Region	*Average price*		*Percent of U.S. average*	
	1960	*1978*	*1960*	*1978*
New England	29.32	64.78	94.0	98.5
North Mid-Atlantic	28.99	63.65	92.9	96.8
South Mid-Atlantic	30.35	64.13	97.3	97.5
Southeast	31.39	64.80	100.6	98.5
Great Lakes	31.07	66.21	99.6	99.8
Southwest	30.21	65.66	96.9	99.8
Central Plains	30.10	66.85	96.5	101.6
Rocky Mountain–Northern Plains	33.92	67.32	108.8	102.3
Far West	31.82	68.02	102.0	103.4
Northwest	34.74	67.73	111.4	103.0
All United States	31.19	65.78	100.0	100.0

Source: American Petroleum Institute, *Basic Petroleum Data Book, Petroleum Industry Statistics*, sec. 4, table 2 (Washington, D.C., API, 1981).

Table 13. *Delivered Cost per Short Ton of Coal (Average Price) to Electric Utilities, by Region, 1966 and 1978*[a]
Dollars

Region	*Average price*		*Percent of U.S. average*	
	1965	*1978*	*1965*	*1978*
New England	9.65	42.50	143.2	214.1
North Mid-Atlantic	7.58	34.34	112.5	173.0
South Mid-Atlantic	7.08	32.60	105.0	164.2
Southeast	5.99	33.48	88.9	168.7
Great Lakes	6.33	25.41	93.9	128.0
Southwest	5.31	14.49	78.8	73.0
Central Plains	8.02	23.37	119.0	117.7
Rocky Mountain–Northern Plains	4.74	12.76	70.3	64.3
Far West	5.65	19.17	83.8	96.6
Northwest	n.a.	11.00	n.a.	55.4
All United States	6.74	19.85	100.0	100.0

Sources: For 1965, U.S. Federal Power Commission, *Delivered Prices, Steam-Electric Plant Construction Cost, and Annual Production Expenses* (GPO, 1967), pp. 1–161; and for 1978, Department of Energy, Energy Information Administration, *Delivered Prices, Steam-Electric Plant Construction Cost and Annual Production Expenses* (GPO, 1980), pp. 3–184.

n.a. Not available.

a. State estimates are average delivered cost per unit to state utilities.

Table 14. *Average Rates for Electric Power, by Region, 1965 and 1978*
Cents per kilowatt hour

Region	Average rate[a]						Percent of U.S. average					
	Residential		Commercial		Industrial		Residential		Commercial		Industrial	
	1965	1978	1965	1978	1965	1978	1965	1978	1965	1978	1965	1978
New England	2.31	4.83	3.65	7.08	1.54	3.47	112.1	118.7	116.6	118.2	117.6	108.8
North Mid-Atlantic	2.39	5.49	3.78	9.42	1.21	5.19	116.0	134.9	120.8	157.5	92.4	162.7
South Mid-Atlantic	2.05	4.33	3.33	6.83	1.32	3.40	99.5	106.4	106.4	114.2	100.8	106.6
Southeast	1.83	3.72	2.92	5.62	1.10	3.05	91.3	91.4	93.3	94.0	84.0	95.6
Great Lakes	1.98	3.84	2.95	5.59	1.50	3.45	96.1	94.3	94.2	93.5	114.5	108.1
Southwest	2.10	4.12	3.42	5.62	1.35	3.14	101.9	100.7	109.3	94.0	103.1	98.4
Central Plains	2.05	3.99	3.41	5.89	1.38	3.11	99.5	98.0	108.9	98.5	105.3	97.5
Rocky Mountain–Northern Plains	2.09	3.49	3.15	5.13	1.32	2.45	101.5	87.7	100.6	85.8	100.8	76.8
Far West	2.25	4.81	2.90	6.42	1.43	3.93	117.0	118.2	92.7	107.4	109.2	123.2
Northwest	1.86	2.93	1.98	4.05	0.89	2.01	74.3	72.5	63.3	67.7	67.9	63.0
All United States	2.08	4.07	3.13	5.98	1.31	3.19	100.0	100.0	100.0	100.0	100.0	100.0

Sources: For 1965, Federal Power Commission, *Typical Electrical Bills* (GPO, 1966); and for 1978, Department of Energy, Energy Information Administration, *Typical Electrical Bills* (GPO, 1979).
Residential: for 1965, table C, p. ix, c = charge, then c/kwh = c/500 kwh; and for 1978, table 4, p. xvii (LB + HB)/(2 × 500 kwh).
Commercial: formula: average of 750 kwh bill for state/750. For 1965, table 2, pp. 96–106; and for 1978, table 10, pp. 147–64.
Industrial: state average bill, billing demands based on monthly consumption of 120,000 kwh. For 1965, table 3, pp. 108–24; and for 1978, table 11, pp. 166–84.
a. Based on a level of 500 kwh for residential use and 750 kwh for commercial use.

The pattern of rates for commercial electricity is much the same. In 1965, they ranged from a low of 63 percent of the U.S. average in the Northwest to almost 121 percent in the North Mid-Atlantic states. The spread had widened substantially by 1978. Commercial users in the Northwest paid 68 percent of the U.S. average in that year, while those in New York and New Jersey paid more than 157 percent. The story is essentially repeated when changes in industrial energy prices are observed, with one significant exception. In 1965, the lowest industrial rates were in the Northwest. The highest were in New England, followed by the Great Lakes region. The latter includes the heavily industrial states of Illinois, Indiana, Michigan, and Ohio. By 1978, while the Northwest remained at the bottom of the list at 63 percent of the U.S. average, the North Mid-Atlantic region had moved to the top. In New York and New Jersey, the price of industrial energy was almost 63 percent higher than the U.S. average in 1978.

The other region to register a significant increase in industrial energy prices in relation to the U.S. average was the Far West. There were smaller increases in the South Mid-Atlantic and the Southeast. In the remaining regions, industrial energy prices dropped in relation to the U.S. average. It is impossible to tell from the available information the extent to which these switches—particularly the large switch in the North Mid-Atlantic region—were caused by economic forces rather than the effects of the regulatory process.

Regional comparisons of natural gas prices are given in table 15. Only one distribution is available, that between residential and industrial users. In 1965, the lowest residential prices were in the Southwest, the Central Plains, and the Rocky Mountain–Northern Plains, where most of the country's natural gas is produced. The highest price, by a substantial margin, was in New England. In 1978, the gas-producing regions remained at the bottom of the list. Residential gas prices in New England, the highest in the country, increased more slowly than the national average. They were still equal to 164 percent of that average in 1978. There was less convergence toward the U.S. average in New York and New Jersey. In general, however, the range of residential gas prices among regions narrowed substantially between 1965 and 1978.

Industrial gas prices, while lower across the board than those charged residential users, follow much the same pattern of regional variation. The range narrowed between 1965 and 1978, but in the northeastern regions and in the Far West, prices remained considerably higher than

Table 15. *Average Gas Utility Prices, by Region, 1965 and 1978*[a]
Dollars per million Btu

					Percent of U.S. average			
	Residential		*Industrial*		*Residential*		*Industrial*	
Region	*1965*	*1978*	*1965*	*1978*	*1965*	*1978*	*1965*	*1978*
New England	2.36	4.15	2.03	3.65	233.7	164.0	327.4	167.4
North Mid-Atlantic	1.52	3.67	1.29	3.37	150.5	145.1	208.1	154.6
South Mid-Atlantic	1.30	3.23	0.98	2.84	128.7	127.7	158.1	130.3
Southeast	1.25	2.64	0.56	2.02	123.8	104.3	90.3	92.7
Great Lakes	1.00	2.40	0.73	2.17	99.0	94.9	117.7	99.5
Southwest	0.75	2.15	0.35	1.74	74.3	85.0	56.5	79.8
Central Plains	0.80	2.03	0.51	1.75	79.2	80.2	82.3	80.3
Rocky Mountain–Northern Plains	0.73	2.06	0.51	1.77	72.3	81.4	82.3	81.2
Far West	1.74	4.09	1.17	3.39	172.3	161.7	188.7	155.5
Northwest	1.39	3.04	0.69	2.31	137.6	120.2	111.3	106.0
All United States	1.01	2.53	0.62	2.18	100.0	100.0	100.0	100.0

Source: American Gas Association, Department of Statistics, *Gas Facts 1979: A Statistical Record of the Gas Utility Industry* (Arlington, Virginia: AGA, 1980), pp. 120–21.
a. Revenues from resale are not included.

the U.S. average. Industrial rates in the Southwest, which includes Louisiana, Oklahoma, and Texas, were the lowest throughout the period.

Distillate oil prices, by sector and region, are given in table 16. This grade of oil was chosen because it is used by residential, commercial, and industrial customers and comparison can therefore be made across sectors as well as among regions.[16] The lighter grades of residential distillate oil consistently sell for more than the commercial and industrial grades.

In 1965, the residents of the Rocky Mountain–Northern Plains, the Far West, and the Northwest paid the highest prices for fuel oil. Prices in New England and the Southeast were close to the U.S. average. In other regions prices were lower than that average, ranging from 91 percent in the Central Plains to almost 99 percent in the North Mid-Atlantic states. The range had narrowed by 1979, by which year prices in New England were higher than the U.S. average. The price in the

16. Distillate fuel oil includes no. 1 and no. 2 heating oils, diesel fuels, and no. 4 fuel oil. It is used for space heating, engine fuel, and the generation of electric power. Residual fuel oil, used for the generation of electric power, space heating, vessel bunkering, and various industrial purposes, consists of no. 5 and no. 6 fuel oils; see U.S. Department of Energy, Energy Information Administration, *Annual Report to Congress, 1978*, vol. 2, DOE/EIA-0173, pp. 166, 170.

Table 16. *Average Prices of Distillate Fuel, by Sector and Region, 1965 and 1979*
Cents per gallon

Region	*Average price*						*Percent of U.S. average*					
	Residential		*Commercial*		*Industrial*		*Residential*		*Commercial*		*Industrial*	
	1965	1979	1965	1979	1965	1979	1965	1979	1965	1979	1965	1979
New England	15.72	75.50	13.49	70.05	13.42	70.05	100.19	103.65	99.34	102.22	99.19	103.85
North Mid-Atlantic	15.48	73.11	13.53	68.82	13.39	68.51	98.66	100.37	99.63	100.42	98.97	101.57
South Mid-Atlantic	14.97	70.38	13.44	67.09	13.34	66.86	95.41	96.62	98.97	97.90	98.60	99.13
Southeast	15.74	70.41	14.05	66.95	13.93	66.71	100.32	96.66	103.46	97.69	102.96	98.90
Great Lakes	14.60	71.92	12.92	68.56	12.86	68.42	93.05	98.74	95.14	100.04	95.05	101.44
Southwest	15.00	72.06	12.82	67.81	12.71	67.55	95.60	98.93	94.40	98.95	93.94	100.15
Central Plains	14.34	71.50	12.89	69.16	12.83	69.06	91.40	98.16	94.92	100.92	94.83	102.39
Rocky Mountain–Northern Plains	16.51	68.12	13.98	65.26	13.96	65.22	105.23	93.52	102.95	95.23	103.18	96.69
Far West	17.60	68.20	14.52	65.04	14.52	65.04	112.17	93.63	106.92	94.91	107.32	96.43
Northwest	17.60	65.96	14.52	62.86	14.52	62.86	112.17	90.55	106.92	91.73	107.32	93.19
All United States	15.69	72.84	13.58	68.53	13.53	67.45	100.00	100.00	100.00	100.00	100.00	100.00

Source: Department of Energy, Energy Information Administration, *State Energy Fuel Prices, by Major Economic Sector, from 1960 through 1979* (GPO, 1981).

Northwest, meanwhile, had dropped to slightly more than 90 percent of the average.

Regional shifts in the prices of commercial distillate oil were similar to those in the residential sector. The high-cost regions in 1965 were the three western regions, although prices in the Southeast were also higher than the U.S. average. By 1979, the price in New England was the highest in the country. The lowest price for commercial distillate in 1965 was in the Southwest, but the lowest price for commercial oil in 1979 was in the Northwest. The spread in commercial oil prices narrowed between 1965 and 1979, but not as much as did the spread in residential fuel oil prices.

There was a somewhat different pattern of change in regional industrial oil prices. The highest prices in the country in 1965 were in the Far West and the Northwest, followed by the Rocky Mountain and Northern Plains states. By 1979, the highest price was in New England, followed by the Central Plains. Meanwhile, industrial oil prices in the Northwest, a beneficiary of the Alaskan oil boom, had dropped below the U.S. average. Regional differences in industrial oil prices narrowed slightly less than in commercial prices during the period covered by table 16.

The patterns of regional price changes are far from uniform among fuels. Retail gasoline prices are the least variable from region to region. Not only were coal prices much higher in 1978 than in 1965, the spread between the lowest and highest prices had widened. The range of prices of electric energy among regions was considerably wider in 1978 than in 1965, although the reverse was true of the prices of gas utilities. The interregional range of oil prices narrowed between 1965 and 1977. The trend toward uniformity was most pronounced in the residential sector, followed by prices in the industrial and commercial sectors, in that order. There were also marked interregional shifts. Regions in which oil prices were highest in 1965 had dropped below the national average by 1977, while oil prices in the northeastern regions moved well above that average. Meanwhile, the position of the Southwest, already favorable in 1965, had improved across the board by 1977. The significance of these changes for regional development will be examined in the following section.

The Differential Effects of Rising Energy Prices

As noted earlier, the basic thesis of this study is that the primary regional effect of rising energy prices has been on the terms of trade

Figure 3. *Hypothetical Exchange Ratios, 1970–78*

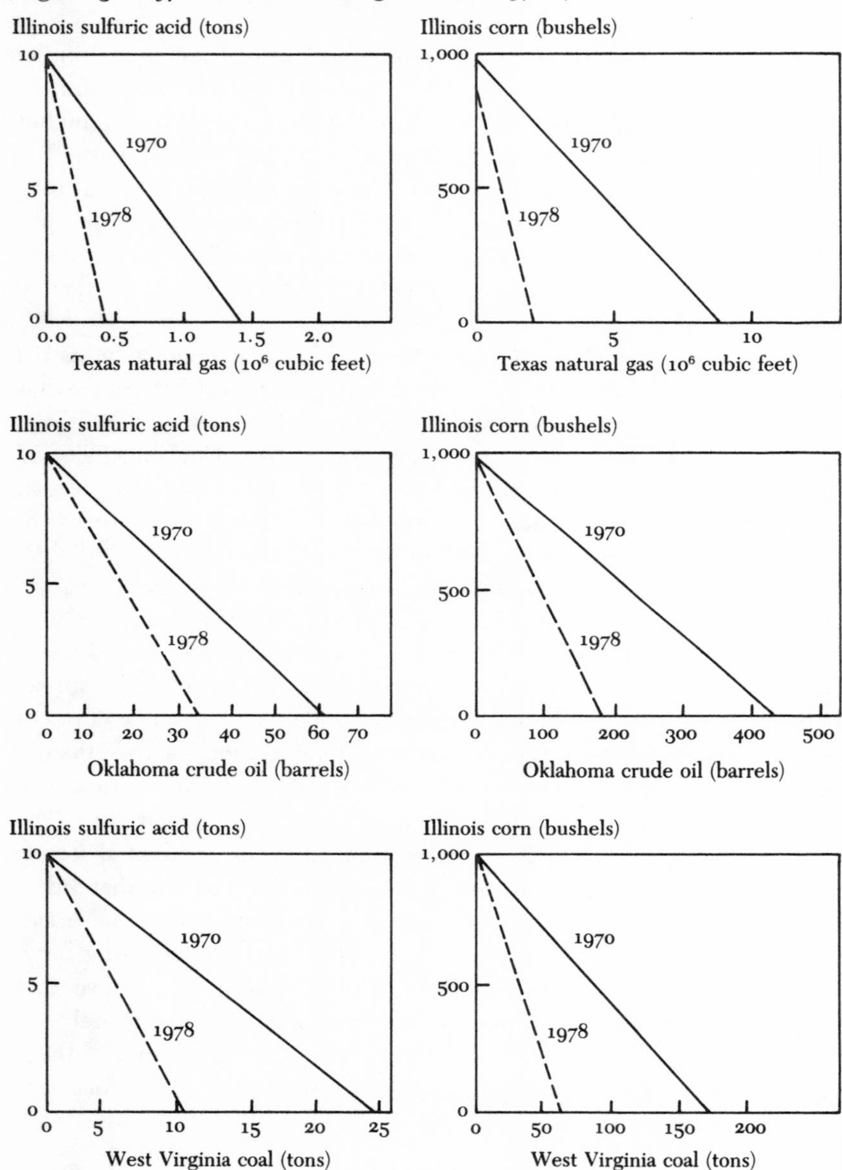

Source: Hans H. Landsberg and Joseph M. Dukert, *High Energy Costs: Uneven, Unfair, Unavoidable?* (Johns Hopkins University Press for Resources for the Future, 1981), p. 94.

between energy-surplus and energy-deficit states rather than on the location of manufacturing activity. Changing terms of trade have, however, caused transfers of income from energy-deficit to energy-surplus states and have thus doubtless accelerated the shift of other economic activities—particularly those that are attracted to markets—from the Northeast to the South and West.[17] Historically, the terms of trade favored industrial states, or those with tertiary sectors serving national or large regional markets. The prices of fossil fuels were stable or declining, while the prices of manufactured goods rose. There was a flow of economic rents from states with fossil fuel export bases to the energy-deficit states. But that situation has changed. Since 1970, the prices of fossil fuels have increased faster than the general price level.[18] The result has been a shift in the terms of trade in favor of fossil fuels. This has been one of the factors contributing to the continuing decline in relative income per capita in energy-deficit states, while the same measure of economic well-being has increased in energy-surplus states.

The sort of shift engendered by increasing fuel prices since 1970 is illustrated by figure 3, which shows exchange ratios between the three fossil fuels and two commodities, one industrial and the other agricultural.[19] The exchange ratios are labeled "hypothetical" only because there is no way of knowing whether the products were actually traded between the states involved. The data, however, are real.

In 1970, 1.4 million cubic feet of Texas natural gas would have exchanged for 10 tons of Illinois sulfuric acid. By 1978, only 400,000 cubic feet of gas would have been required to pay for 10 tons of sulfuric acid. Similarly, 61 barrels of Oklahoma oil would have exchanged for 10

17. There has been a similar shift in the terms of trade between older industrial states and some agricultural states, although that shift has been less clear-cut because of cyclical influences on farm prices and production; see Miernyk, "Regional Shifts in Economic Base and Structure in the United States since 1940," in Victor L. Arnold, ed., *Alternatives to Confrontation: A National Policy toward Regional Change* (D. C. Heath, 1980), pp. 106–20.

18. For a graphic illustration, see Diane DeVaul, *National and State Energy Expenditures, 1970–1980* (Washington, D.C.: Northeast-Midwest Institute, 1981), p. 4.

19. The two cases in figure 3 were selected from a set of fifteen such comparisons made in an earlier study. All the shifts were in the same direction, whether the commodity was a raw material, such as corn, a processed agricultural product, such as butter, or an industrial product, such as sulfuric acid. To be significant, such comparisons must be limited to homogeneous products or services. For the complete set of comparisons and sources of data, see Hans H. Landsberg and Joseph M. Dukert, *High Energy Costs: Uneven, Unfair, Unavoidable?* (Johns Hopkins University Press for Resources for the Future, 1981), table A-2, p. 94.

Figure 4. *Changes in Relative per Capita Income and the Price of Electric Energy, Selected States, 1973–79*

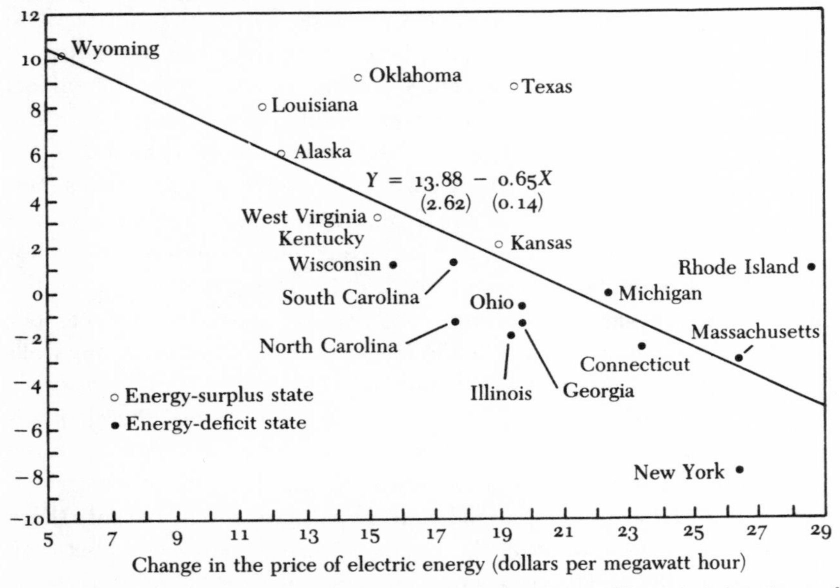

Sources: Author's calculations from data in Department of Commerce, Bureau of Economic Analysis, *Survey of Current Business*, August 1974 and August 1980, table 2; Edison Electric Institute, *Statistical Yearbook of the Electric Utility Industry, 1974* (Washington, D.C.: EEI, 1975); and ibid., *1980* (EEI, 1981).

tons of sulfuric acid in 1970, but only 32 barrels of oil would have been required in 1978. Finally, it would have required more than 24 tons of West Virginia coal to purchase 10 tons of Illinois sulfuric acid in 1970. The same transaction would have required only 10 tons of coal in 1978. The corresponding shifts in the ratios of Illinois corn to the three fossil fuels are shown on the right side of figure 3.

The relations between changes in selected energy prices and changes in per capita income have been examined by means of simple regression analysis. The observations are a group of energy-surplus and energy-deficit states.

Figure 4 shows changes in per capita income as a percentage of the U.S. average and changes in the prices of electric energy between 1973 and 1979. The coefficient of correlation is 0.57, and the F-statistic is significant at the 0.01 level. This is not a particularly good statistical fit,

Figure 5. *Changes in Relative per Capita Income and the Price of Natural Gas, Selected States, 1973–79*

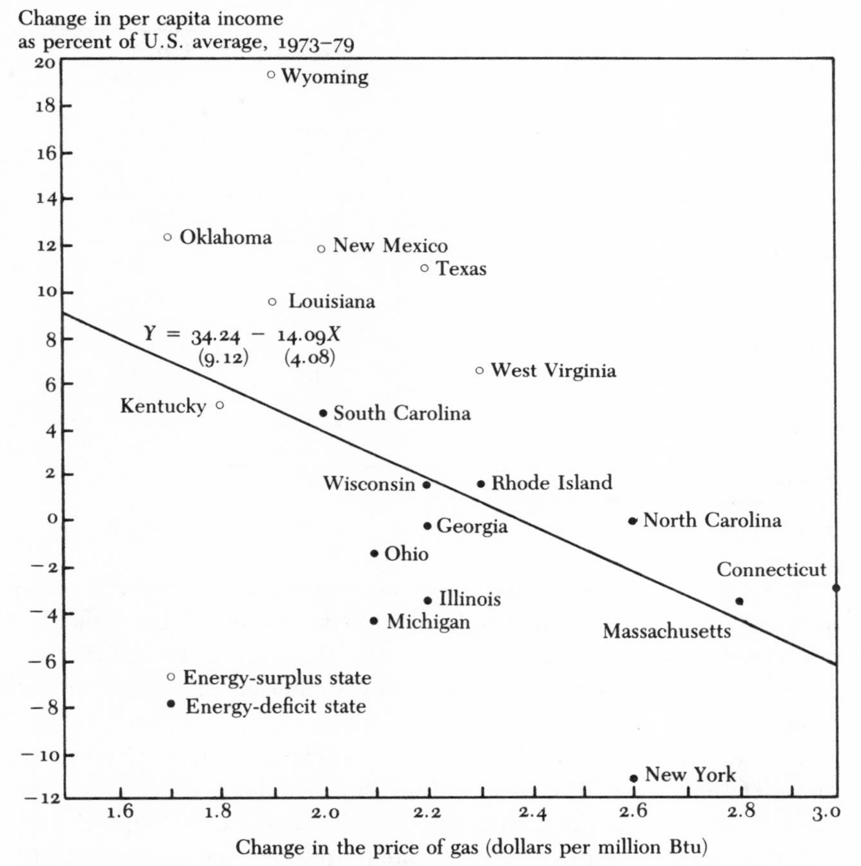

Sources: Author's calculations from data in *Survey of Current Business*, August 1974 and August 1980, table 2; American Gas Association, *Gas Facts*, 1974; and ibid., 1980.

but the scatter diagram illustrates in a general way the shifts discussed in the text.

The second scatter diagram (figure 5) relates changes in relative income per capita to changes in natural gas prices during the same period. The coefficient of correlation in this case is 0.431, and the F-statistic is again significant at the 0.0031 level. As in the first diagram, the energy-surplus states are bunched in the upper-left quadrant, while the energy-deficit states are in the lower-right quadrant.

Table 17. *Relation between Energy Self-Sufficiency, 1978–79, and Changes in Relative Income per Capita, 1959–79, by Standard Federal Region*

Region	*Energy production minus energy consumption (as percent of U.S. totals)*[a]	*Percent change in relative income per capita, 1959–79*
New England	−3.99	−6.6
North Mid-Atlantic	−7.15	−13.7
South Mid-Atlantic	0.45	0.8
Southeast	−6.07	18.9
Great Lakes	−15.00	−2.9
Southwest	29.51	13.3
Central Plains	−3.43	3.8
Rocky Mountain–Northern Plains	6.85	5.6
Far West	−4.77	−6.1
Northwest	3.62	1.2

Sources: Tables 3, 4, and 5.
a. Data on consumption are for 1978; data on production are for 1979.

The same method was applied to changes in energy prices and changes in the ratio of manufacturing employment to population.[20] The coefficients of correlation were 0.2, for electric energy, and 0.018, for natural gas. Neither of the regressions is statistically significant at any acceptable level.

The results of these regressions are not offered as tests of the basic thesis of this paper. The observations, for one thing, are not random samples. Figures 4 and 5 are consistent, however, with the argument that rising energy prices have had differential regional income effects. The lack of correlation between changes in energy prices and manufacturing employment by state is also consistent with the hypothesis that differential changes in energy prices have had relatively little effect on the location of industry.

Changes in relative income per capita—that is, changes in regional income per capita as a percentage of the U.S. average—and the energy-deficit or surplus position of the SFRs are compared in table 17. The

20. Manufacturing employment was divided by population to eliminate scale effects. During the relatively short period involved, this should have had no appreciable effect on the regressions.

figures in this table were derived from earlier tables on regional production, consumption, and per capita income.

Except in two regions—the Southeast and the Central Plains—energy deficits have been accompanied by declines in relative income per capita. No attempt will be made here to explain why a region with a moderate energy deficit should have the largest increase in relative per capita income.[21] But with this exception, and a smaller discrepancy in the Central Plains, there is a fairly close correspondence between energy self-sufficiency and both the direction and the magnitude of changes in regional income per capita.

In the study from which figure 3 was derived, employment in each region was classified on the basis of sensitivity to changes in energy prices.[22] This classification was based on an earlier study in which a 480-order national input-output table was used to rank activities, at the four-digit level, on the basis of their energy intensity. The inverse matrix was used to calculate both direct and indirect energy requirements per dollar of output. The sectors were then classified by identifying the factor or factors primarily responsible for the location of establishments. Some, such as blast furnaces and basic steel mills, clearly try to minimize transport costs. Others—including, for example, the cement, brick, and structural clay sectors—must consider both resources and transport costs when choosing a location. Extractive activities, which include the production of all fossil fuels, are tied completely to resources. But activities covering a wide range are not tied to materials or markets and are not likely to be influenced by transport costs.[23] Such activities could be sensitive to changes in energy prices. Those in this group that serve national or broad regional markets were classified as directly vulnerable activities. Others that are suppliers to the first group were considered to

21. One contributing factor is that the Southeast started from the lowest income base in 1959; for additional discussion see Miernyk, "The Changing Structure of the Southern Economy," in E. Blaine Liner and Lawrence K. Lynch, eds., *The Economics of Southern Growth* (Research Triangle Park, N.C.: Southern Growth Policies Board, 1977), pp. 35–63. In a paper by David Manuel, "The Effects of Higher Energy Prices on State Income Growth," *Growth and Change*, vol. 13 (July 1982), pp. 26–37, the income-shift hypothesis advanced here is supported.

22. William Miernyk, "The Differential Effects of Rising Energy Prices on Regional Income and Employment," in Hans Landsberg, ed., *High Energy Costs: Assessing the Burden* (Johns Hopkins University Press for Resources for the Future, 1982), pp. 297–330.

23. The details of this classification are given in Miernyk and others, *Regional Impacts*, pp. 82–84.

Table 18. *Percentage Distribution of U.S. Employment, by Sensitivity to Changes in Energy Prices and by Region, 1970 and 1977*

	Directly vulnerable		*Indirectly vulnerable*		*Sheltered*		*Total*	
Region	*1970*	*1977*	*1970*	*1977*	*1970*	*1977*	*1970*	*1977*
New England	7.1	6.5	5.2	4.9	6.7	6.2	6.8	6.2
North Mid-Atlantic	12.0	9.7	17.4	13.2	14.9	12.4	14.4	12.0
South Mid-Atlantic	12.5	11.4	11.2	10.8	11.4	10.9	11.6	11.0
Southeast	16.2	17.8	13.2	14.5	13.9	15.0	14.3	15.4
Great Lakes	29.3	28.0	20.0	19.8	21.6	21.0	23.1	22.2
Southwest	6.6	8.4	10.7	12.0	9.2	10.4	8.7	10.1
Central Plains	4.3	4.8	5.3	5.6	5.4	5.3	5.2	5.2
Rocky Mountain–Northern Plains	1.2	1.5	2.4	2.9	2.4	2.9	2.1	2.7
Far West	8.1	9.0	11.4	12.5	11.8	12.7	11.0	12.0
Northwest	2.5	3.0	3.0	3.4	2.8	3.2	2.7	3.1
All United States	100.0	100.0	100.0	100.0	100.0	100.0	100.0	100.0

Source: William Miernyk, "The Differential Effects of Rising Energy Prices on Regional Income and Employment," in Hans H. Landsberg, ed., *High Energy Costs: Assessing the Burden* (Johns Hopkins University Press for Resources for the Future, 1982), pp. 315–16. Figures are rounded.

be indirectly vulnerable. The remainder—those that are primarily concerned with resources, are primarily concerned with transport, or are strictly local-market activities—were considered to be sheltered from the direct regional effects of changes in energy prices.

On the basis of this classification, about 21 percent of U.S. employment in 1970 was in directly vulnerable sectors, with slightly less than 5 percent in the indirectly vulnerable sector. The remainder—almost 75 percent—was in sectors considered to be sheltered from the direct effects of differential changes in energy prices.

The regional distribution of employment, by these categories, in 1970 and 1977, is given in table 18. As a benchmark, the distribution of total employment, by SFR, is given in the right-hand column. There was an appreciable decline in employment in both directly and indirectly vulnerable activities in the North Mid-Atlantic, an energy-deficit region, but there was also a pronounced drop in sheltered employment in that region during the same period. On the other hand, employment in all three categories increased in the Southwest, the region with the largest energy surplus. Such across-the-board changes—one negative and one positive—show how difficult, if not impossible, it is to identify locational determinants using aggregate data.

It can be said, however, that some of the decline in employment in New York and New Jersey can be attributed to rapid increases in energy prices in that region and that some of the increase in employment in the Southwest can be attributed to favorable energy prices there. The regional losses and gains in sheltered activities—which in the country as a whole are three times as large as in the vulnerable sectors—cannot be directly attributed to either unfavorable or favorable changes in regional prices of energy.

Part of the explanation of interregional shifts in sheltered employment can be deduced from a study by Giarratani and Socher. Using conventional location theory and an interindustry supply model, they considered an aspect of the energy-location issue that had previously been ignored. A substantial—and growing—share of economic activity in the United States is being drawn toward its markets. If population continues to shift to the West and the Southwest, as is likely, and if the per capita income of energy-surplus states rises faster than it does in energy-deficit states, these activities will be attracted to the energy-surplus states, not by energy prices or availability per se, but by growing markets, engendered by rising energy prices.[24] Some shifts in activities toward their markets may be considered to be energy-induced. While it is true that long-run shifts in population and income go back to the beginning of the country, the relocation of activities nearer their markets has probably been accelerated by differential changes in regional energy prices.

What about offsetting tendencies? What will be the long-run effects on industrial location of district utility systems, cogeneration, and other approaches to conservation that might generate economies sufficient to reduce the importance of energy as a locational determinant?[25] There is evidence that rising energy prices have already stimulated conservation efforts, such as better insulation and the elimination of direct waste heat. But U.S. industry has a long way to go to catch up with other countries

24. For further details, see Frank Giarratani and Charles F. Socher, "The Pattern of Industrial Location and Rising Energy Prices," in Miernyk and others, eds., *Regional Impacts*, pp. 103–16.

25. For a discussion of cogeneration and its potential effect on some energy-intensive industries, see H. S. Burness, R. G. Cummings, and I. Paik, "Environmental Trade-offs with the Adoption of Cogeneration Technologies," *Journal of Environmental Economics and Management*, vol. 8 (March 1981), pp. 45–58.

Table 19. *Differences in the Price of Coal between West Virginia and Selected Destinations, 1965, 1971, and 1976*
Prices in dollars per ton

Year	*Price f.o.b. West Virginia*	*Philadelphia, Pa.*			*Trenton, N.J.*			*Rochester, N.Y.*			*Merrimack, N.H.*		
		Delivered price	*Difference*		*Delivered price*	*Difference*		*Delivered price*	*Difference*		*Delivered price*	*Difference*	
			Price	*Percent*		*Price*	*Percent*		*Price*	*Percent*		*Price*	*Percent*
1965	5.00	7.96	2.96	37.2	7.51	2.51	33.4	8.05	3.05	37.9	9.52	4.52	47.4
1971	8.07	13.16	5.09	38.7	13.46	5.39	40.0	13.47	5.40	40.1	11.82	3.75	31.7
1976	30.60	33.73	3.13	9.3	42.05	11.45	27.2	33.01	2.41	7.3	34.33	3.73	10.9

Sources: Prices f.o.b. West Virginia, U.S. Bureau of Mines, *Minerals Yearbook 1965*, vol. 2, *Mineral Fuels* (GPO, 1966), p. 182; ibid., 1971, vol. 1, *Metals, Minerals, and Fuels*, p. 371; ibid., 1976, vol. 1, *Metals, Minerals, and Fuels* (GPO, 1978), p. 352; delivered prices, Department of Energy, *Delivered Prices, Steam-Electric Plant Construction Cost, and Annual Production Expenses*, 1965, pp. 85–117; ibid., 1971, pp. 85–119; ibid., 1976, pp. 97–137.

that have never enjoyed the luxury of "cheap" energy. Continued increases in energy prices are almost certain to stimulate more sophisticated conservation measures. The possible effects of such measures on the stabilization of patterns of industrial location are still largely conjectural. There is no reason to believe that cogeneration and other conservation techniques would be practiced only in energy-deficit states. Once the technology became available, there would be as much incentive to use the same techniques in energy-surplus states as in energy-deficit states. It is clear, of course, that effective conservation measures would reduce the *relative* cost of energy, and thus reduce the importance of interregional differentials as potential determinants of location.

Another possibility has been suggested by Milne, Glickman, and Adams, who used a multiregional econometric model to simulate the regional effects of changing energy prices.[26] They argued, among other things, that as the price of energy continues to rise "the transportation differential between regions becomes less important because it constitutes a smaller proportion of the total delivered cost of the energy product."[27] There is some fragmentary evidence to support this hypothesis (see table 19).

The data in table 19 must be interpreted cautiously, however. First, the f.o.b. prices shown are averages for the state of West Virginia. This doubtless accounts for some of the unusual variation in delivered prices among the four destinations. There is no other way to explain why it cost less to deliver coal from West Virginia to Merrimack, New Hampshire, than to Philadelphia, since the former is 505 miles from Morgantown, West Virginia—a major shipping point 75 miles south of Pittsburgh—while Philadelphia is only 253 miles from Morgantown. But even if the figures in table 19 are taken with a large grain of salt, they support the hypothesis advanced by Milne and his associates.

Between 1965 and 1971, relative transportation costs increased in all instances but one. Between 1971 and 1976, however, when the price of coal rose dramatically, the drop in relative transportation costs was substantial. This might reflect, to some extent, regulatory lag, since increases in freight rates had to be approved by the Interstate Commerce Commission. But even if transportation costs had increased substantially

26. William J. Milne, Norman J. Glickman, and F. Gerard Adams, "A Framework for Analyzing Regional Growth and Decline: A Multiregional Econometric Model of the United States," *Journal of Regional Science*, vol. 20 (May 1980), pp. 173–89.

27. Ibid., p. 186.

more than they did, the relative differentials in delivered prices would probably still have declined significantly.

It is possible that both these tendencies could prevail in the future. Energy-intensive and some market-oriented activities could continue to be attracted to energy-surplus states. On the other hand, regional differentials in delivered prices of energy could be narrowed, as suggested by table 19, if transportation costs rose more slowly than wellhead or mine-mouth fuel prices.

Another factor that might influence migration rates—both of population and of economic activity—is the effect of rising transportation costs on commuting. The newer cities of the South and West are far more sprawling than the older cities of the Northeast, and many are less well served by public transportation systems. The lower housing densities in these regions, with correspondingly longer commuting distances, could make them somewhat less attractive as energy prices rise. Eventually, the structures of the newer cities will change to accommodate rising transportation costs, but this change will be slow in coming about. New residents will doubtless try to locate closer to their jobs in the future than they have in the past, but an increase in the demand for housing in inner cities will drive prices up. Only speculation about the potential effects of such changes on migration rates is possible at this stage, but it is likely that they will tend to slow them down.

What actually happens in the long run will depend largely on the national economic growth rate and public policy toward regional energy-price differentials. Before the effects of the national growth rate on regional growth are discussed, however, a few words should be said about the question of state severance taxes on fossil fuels.

Taxes on Nonrenewable Energy Resources

A feeling appears to be growing among public officials in energy-deficit states that these states will become the victims of "exorbitant" severance taxes and other taxes on fossil fuels purchased from energy-surplus states. A recent expression of this concern is found in a pamphlet with the colorful title *The United American Emirates*.[28]

28. Tom Cochran and J. R. Prestige, *The United American Emirates* (Washington, D.C.: Northeast-Midwest Institute, 1981).

Table 20. *Energy-Surplus States, by Rank and Tax Rates, 1978*
Percent of value

	Tax rate		
Rank in exports	*Oil*	*Gas*	*Coal*
1. Louisiana	12.5	3.5[a]	0.2
2. Wyoming	4.0	4.0	10.5
3. New Mexico	7.0	5.8	8.4
4. Kentucky	1.5	n.a.	4.5
5. Alaska	12.25	10.0	n.a.
6. Oklahoma	7.0	7.0	0.3
7. West Virginia	4.35	8.63	3.5
8. Montana	2.1–2.65	2.65	30.5
9. Texas	4.6	7.5	n.a.

Sources: Rank in exports from Chase Econometrics, cited by Richard Corrigan and Rochelle L. Stanfield, "Rising Energy Prices—What's Good for Some States Is Bad for Others," *National Journal*, March 22, 1980, p. 469; and tax rates from Tom Cochran and J. R. Prestige, *The United American Emirates* (Washington, D.C.: Northeast-Midwest Institute, 1981), p. 4.
n.a. Not available.
a. Calculated by Harley T. Duncan, staff associate, National Governors Association, March 6, 1980; from Bureau of the Census, *State Government Tax Collections in 1978* (GPO, 1979).

In the discussion that follows, the term "severance tax" will be applied to all state taxes levied on the sale of fossil fuels.[29] Table 20, taken from *The United American Emirates*, lists all but one of the fossil fuel severance taxes levied in 1978. The tax on oil ranged from a low of 1.5 percent in Kentucky to 12.5 percent in Louisiana. The rate in Alaska, 12.25 percent, was just slightly less than that in Louisiana. The tax on natural gas ranged from 2.65 percent in Montana to 10 percent in Alaska. The widest range was in taxes levied on coal. This went from a low of 0.2 percent in Louisiana, where little coal is produced, to 30.5 percent in Montana, where production of coal is expanding.[30]

How much of the revenue of the various states is derived from these taxes? Kentucky, the number one coal-producing state, derived less than 7 percent of its revenue from a relatively modest tax on coal. Louisiana, a major producer of oil and natural gas, derived more than 23 percent of its revenue from severance taxes. The relative importance of severance

29. Not all states label their taxes this way. West Virginia, for example, has a transactions tax, called the Business and Occupations Tax, which is levied on all sectors, including fossil fuels. The rates vary from sector to sector.

30. The report by Cochran and Prestige did not include the North Dakota coal severance tax. That tax was first levied in 1975 and has twice been amended. The rate, as of November 1981, is $1.00 a ton, with a one-cent increase for every four-point rise in the wholesale price index of the U.S. Department of Labor. This information was given me by David Haring of the North Dakota State Tax Department, November 11, 1981.

Table 21. *Severance Taxes as Percentages of State Revenues, 1978*

State	*Percent*
Louisiana	23.55
Wyoming	22.81
New Mexico	19.16
Kentucky	6.96
Alaska	19.13
Oklahoma	17.51
West Virginia	13.99[a]
Montana	11.95
Texas	17.68

Source: Cochran and Prestige, *United American Emirates*, p. 5.
a. Calculated from Bureau of the Census, *State Government Tax Collections in 1978* (GPO, 1979), and data supplied by the state.

taxes to other producers and exporters of basic energy is presented in table 21.

The severance-tax issue, already controversial, is likely to become even more so in the future. Energy-deficit states, such as those that have banded together to establish the Northeast-Midwest Coalition, are already attempting to impose limits on state severance taxes. But representatives of energy-surplus states are banding together to oppose federal intervention. One group of utilities attempted to have the 30 percent severance tax in Montana declared unconstitutional, but the Supreme Court ruled against it. The energy-deficit coalition has now shifted its efforts to Congress, hoping to achieve by way of legislation what it could not achieve through the courts.

Representatives of states that levy severance taxes have been anything but passive about the efforts of the Northeast-Midwest Coalition to limit such taxes. The Western Governors' Policy Office, which speaks for the governors of the Mountain states, the High Plains states, and Alaska, has met "to plan a more aggressive strategy for opposing limits on severance taxes."[31] They will have the active support of most of the governors of the energy-producing Appalachian states.

It is not likely that state severance taxes will be limited by federal legislation or that state taxes will be replaced by a federal severance tax, "the revenues of which," the Northeast-Midwest Coalition has suggested, "could be shared in some equitable fashion around the nation."[32] Thirty-

31. William E. Schmidt, "Western States Gird to Fight Attempts to Limit Coal Tax," *New York Times*, June 28, 1981.

32. Cochran and Prestige, *United American Emirates*, p. 21.

three states levied some form of severance tax on mineral production in 1980.[33] Most of their lawmakers would doubtless join the Western Governors' Policy Office and the governors of other energy-exporting states in their opposition to efforts to limit state severance taxes.

Conclusions and Future Prospects

The problem of "regional shares" of economic activity and income was exacerbated—as were all other economic problems—by a declining national growth rate during the 1970s. As noted earlier, most regions grow under conditions of robust national growth. Some grow faster than others, and there is a tendency toward convergence of factor returns, as evidenced by changes in relative income per capita. But when the national economy grows slowly, as it has since the early 1970s, some regions grow at the expense of others. Whatever the direction of causality between regional and national growth rates, the two are inseparable.

In his detailed and meticulous analysis of the rate of growth of national income per person employed, Denison found that about a third of the decline in the rate of growth is "explained" by changes in conventional factor inputs, but about two thirds is a statistical residual. Even more startling is the following:

> According to my estimates there is no unexplained drop in productivity until 1974. I consider this timing an important clue in any attempt to unravel the mystery surrounding the retardation. . . . Of course, "coming events cast their shadows before" and the onset of fundamental changes that were to lead to decline may have been discernable in advance of actual events. *But the unexplained decline itelf does not appear until 1974.*[34]

Perhaps Denison's "mystery" is no mystery at all, but a consequence of structural change and the classical law of diminishing returns. If the decline in productivity is a consequence of diminishing returns, it is a natural, irreversible phenomenon, and the trends of the past decade are

33. Steven D. Gold, "State Severance Tax Revenue in 1980," National Conference of State Legislatures, Legislature Finance Paper no. 7 (Denver, 1981), p. 6. The list, incidentally, includes two states, Minnesota and Michigan, that are part of the Northeast-Midwest Coalition.

34. Denison, *Accounting for Slower Economic Growth*, p. 122 (emphasis added).

likely to continue.[35] The result will be further switches in the relative income per capita of energy-exporting and energy-importing states.[36]

If the trends of the past decade continue, will the older industrial, energy-deficient states become economic wastelands? Will they be the depressed areas of the future? There is no reason to believe they will. As noted earlier, about three fourths of the jobs in the United States are in activities that are not directly vulnerable to rising energy prices. There is some variation from region to region, but the range is fairly narrow, from 70 percent in the Great Lakes region to 83 percent in the Rocky Mountain–Northern Plains region.

Rising energy prices are not likely to have a locational effect on most activities—financial, service, educational—that serve national markets from their locations in the Northeast. And few energy-intensive manufacturing activities were located in states in which energy costs were relatively high even before energy prices started to climb steeply during the early 1970s. Older industrial regions will adjust to lower average incomes and modest declines in population and employment as the interregional redistribution of income that has been going on at least since 1929 continues. Even if there is further switching of relative income positions between energy-exporting and energy-importing states, as now seems likely, future divergences from the national average are not likely to be so large as they were a half-century ago.

35. The "law" of diminishing returns should not be confused with economies of scale. It describes the decline in physical output, even with increasing inputs, when at least one factor of production is fixed. In this context, the "fixed" factor is the world's dowry of nonrenewable energy and natural resources. For a definitive statement of the classical law of diminishing returns, see Alfred Marshall, *Principles of Economics: An Introductory Volume*, 8th ed. (New York: Macmillan, 1948), pp. 150–56. For elaboration of the view that constraints on energy and resources affect both national and regional rates of growth as well as regional shares of income and employment, see Miernyk and others, *Regional Impacts*, pp. 40–41 and the references cited on p. 55. See also Miernyk, "Regional Shifts in Economic Base and Structure"; and Miernyk, "Rising Energy Prices and Regional Economic Development," *Growth and Change*, vol. 10 (July 1977), pp. 2–7.

36. In 1970 Alaska was the only state that is now a net exporter of energy in which per capita income was above the average, although it was not at that time an important exporter of energy. A news release of the U.S. Department of Commerce, Bureau of Economic Analysis, "1980 State per Capita Personal Income," dated May 3, 1981, shows that Kansas, Texas, and Wyoming moved into the above-average per capita income class in 1980. Meanwhile, the remaining exporters of energy moved closer to the average. As table 3 shows, a comparison of changes in state per capita incomes understates the extent of the changes in real state per capita incomes. For a set of regional per capita income projections to 2010 and the assumptions on which they are based, see Miernyk, *Regional Analysis and Regional Policy* (Oelgeschlager, Gunn, and Hain, 1982), pp. 101–05.

There is no doubt that differential increases in energy prices have contributed to interregional shifts in population, income, and economic activity. There is a good chance, however, that their influence on the relocation of manufacturing establishments has been exaggerated. The principal effect has been on rates of change in per capita income that reflect shifts in terms of trade between energy-surplus and energy-deficit states.[37] Those shifts have been on the whole orderly, and they represent, essentially, a modest acceleration of long-term trends. The resultant effects on older industrial areas have been less disruptive, for example, than the large-scale, rapid migration of the textile and related industries from the North to the South after World War II. The most reasonable prognosis that can be made at present is that the regions adversely affected by rising energy prices will adjust to new conditions in the future as they adjusted earlier to other changes. They will do so, however, at lower levels of real per capita income than they enjoyed in the past.

Comment by Frank Hopkins

The paper by William Miernyk is a well-documented exposition of the thesis that the principal effect of rising energy prices on regional development is through shifts in the terms of trade between energy-surplus and energy-deficit regions (hypothesis 1) rather than through the price effects that influence industrial location (hypothesis 2).

The structure of the analysis provides a simple yet effective preliminary procedure for analyzing regional effects of factor price changes that could be extended to other variables, including regulations and the availability of transportation service. The initial empirical evidence presented in support of the author's thesis could be strengthened by an examination of the null hypothesis that industries have not relocated in significant numbers because of energy prices. Use of time-series data would also have strengthened the empirical analysis.

37. In an interesting study Lakshmanan expresses agreement with the terms-of-trade hypothesis of this paper. He concludes on the optimistic note, however, that a "region like New England is increasingly specializing in high-valued goods and services that have a high income 'elasticity' of demand (e.g., educational, medical, and high-technology services), leading the transition of the United States into a post industrial services-oriented economy." See T. R. Lakshmanan, "Regional Growth and Energy Determinants: Implications for the Future," *Energy Journal*, vol. 2 (April 1981), p. 22. The income trends in table 2 of this paper and the estimates of "real" income, by region, offer little support for such optimism.

I shall comment on only those points that seem to me either unclear or important enough to the main theme of the paper to be worth restating. Most of my comments are positive, since I agree with the author's principal conclusions.

Introduction

Rejection of hypothesis 2 implies different regional public policy from that implied by acceptance of hypothesis 1. The author stresses that acceptance of hypothesis 2 is inconsistent with traditional industrial location theory. The accepted location theory has extended the microeconomic theory of the firm by including transportation cost, and therefore the industrial-location decision, in the profit-maximizing decision of the firm. Thus, energy is only one factor that influences industrial location, and its effect is restricted to those industries in which energy is a significant factor of production. Rising energy prices would have little effect on service industries that are attracted to their markets, which are beginning to dominate the U.S. economy. Thus, acceptance of the industrial migration hypothesis implies that at the margin rising energy costs dominate all other factors of production. While Miernyk minimizes the importance of energy prices for industrial location, he does recognize that the energy situation may have hastened the decline of industries in the Northeast.

Regional Definition

This section was written to justify the use of the federal census regions as the units of analysis and fulfills a traditional function in regional economics. Unfortunately, its position following the introduction interrupts the flow of the paper. The primary justification for the use of the federal regions appears to be that presentation of data at a lower level of regional disaggregation would be hampered by limitations of the available data and by space constraints.

The Production and Consumption of Energy, by Region

This section could have been improved by the inclusion of an earlier year so that a time-series comparison analysis of changing patterns of production and consumption of energy could be discussed. The time-series comparison would also make the structure of this section consistent with that of the preceding section and would strengthen the evidence used in supporting the theoretical arguments presented later in the paper. I presume that the unavailability of a consistent data series for an earlier year was the primary reason for this omission.

The most useful data in this section are those presented in figure 2, which shows energy surpluses and deficits by region. This figure illustrates the great regional disparities in this measure. The Great Lakes region produces only 85 percent of the total amount of energy it consumes, while the Southwest produces 130 percent of the amount that it consumes. The differences are even more striking at the state level, as shown in table 11.

The Differential Effects of Rising Energy Prices

This is the most important section of the paper. The data presented in earlier sections are used convincingly to outline a strong defense of the terms-of-trade hypothesis. The direct evidence is initiated by some very clever diagrams, which illustrate changes in the terms of trade between two important commodities produced in Illinois and three sources of energy produced in Texas, Oklahoma, and West Virginia. In each case presented, the terms of trade shifted drastically to the energy-producing states between 1970 and 1978. The author notes the hypothetical nature of this argument, since empirical data on trade between the regions illustrated in the graphs do not exist.

With both direct and supporting evidence Miernyk presents convincing qualitative arguments as to the persuasiveness of the terms-of-trade argument over the industrial-location hypothesis. The argument, however, would have been stronger if it could be demonstrated that energy prices have had only a negligible influence on relocation of individual industries that are sensitive to energy prices and that the indirect effects of this industrial change are not significant.

Miernyk has presented a strong theoretical basis for undertaking a quantitative analysis of the regional effects of rising energy prices. Battelle Pacific Northwest Labs and my staff at ORI, Inc., are now engaged in construction of the Metropolitan and State Economic Regional (MASTER) model to complete this task. Fortunately, Professor Miernyk has provided a strong foundation for this work.

The last part of this section has to do with the effects of rising energy prices upon commuting and therefore upon urban form. Rising commuting costs will tend to stimulate the growth of public transportation and to control urban sprawl. Miernyk notes that this process will be slow; recent increases in public transportation and the renewal of the central cities during the last decade, however, indicate that it is already occurring. The financing of public transportation systems in the future may be a significant function of local and state tax revenue. Thus, if the terms of trade for the Northeast and the Midwest continue to increase, the further redevelopment of these areas may be in jeopardy, while the Southwest and the Far West may be able to adjust and plan their growth for greater population densities.

ORI has recently completed, for the National Commission on Air Quality, a study of the potential for the use of coal under alternative regulatory and economic projections. One result of this study was to show the sensitivity of regional coal production to differences in transportation cost, utility plant retirement rates, economic growth rates, emission standards for sulfur dioxide, and the availability of nuclear energy. Our analysis has suggested that the future production share between western and eastern coal is not a critical function of the average national price of coal. Thus, the redevelopment of the cities of the Northeast could be assisted by continued growth of a coal-based industry, and as the industrial sector switches from oil to coal the recent growth of an oil-based western economy may not continue at so rapid a rate as that at which it grew in previous decades.

Taxes on Nonrenewable Energy Resources

One implication of the terms-of-trade hypothesis is that state governments will extract rents from their energy resources to finance their development. Miernyk concludes that for political reasons this process will continue. There are indirect political techniques, however, for using the legislative process to foster the development of one region over others. Specifically, the present requirement that scrubbers be installed on new boilers has favored eastern coal over midwestern and low-sulfur western coal. Other political factors than rising energy prices may have more important implications for regional development and urban form.

Conference Participants

with their affiliations at the time of the conference

Timothy A. Ballard *Technology & Economics*
Herman Bluestone *U.S. Department of Agriculture*
Benjamin F. Bobo *U.S. Department of Housing and Urban Development*
Katharine L. Bradbury *Federal Reserve Bank of Boston*
Dennis W. Carlton *University of Chicago*
Joel Darmstadter *Resources for the Future*
Diane DeVaul *Northeast-Midwest Institute*
Dianne E. Dorius *U.S. House of Representatives*
Anthony Downs *Brookings Institution*
David Engel *U.S. Department of Housing and Urban Development*
Steven Ferrey *National Consumer Law Center*
Bernard J. Frieden *Massachusetts Institute of Technology*
Allen C. Goodman *Johns Hopkins University*
Robert Groberg *U.S. Department of Housing and Urban Development*
Frank Hopkins *ORI, Inc.*
Dwight M. Jaffee *Princeton University*
Ronald Kampe *U.S. Department of Agriculture*
Richard Krashevski *AFL-CIO*
Henry Lee *Harvard University*
Kyu Sik Lee *World Bank*
Dennis L. Little *Library of Congress*
William Marcuse *U.S. Department of Energy*
Jane Meyer *Super-Insulated Building*
William H. Miernyk *University of West Virginia*

Richard D. Morgenstern *Urban Institute*
Terry H. Morlan *U.S. Department of Energy*
Michael P. Murray *Rand Corporation*
Richard F. Muth *Stanford University*
Nancy C. Naismith *U.S. House of Representatives*
Kevin Neels *Rand Corporation*
Frank Parente *AFL-CIO*
Joseph A. Pechman *Brookings Institution*
Mindaugas Petrulis *U.S. Department of Agriculture*
Mary Proctor *U.S. House of Representatives*
John M. Quigley *University of California*
Arthur J. Reiger *U.S. Department of Housing and Urban Development*
Sandra M. Rennie *Mellon Institute*
Elizabeth A. Roistacher *Queens College*
Arthur H. Rosenfeld *University of California (Berkeley)*
Jerome H. Rothenberg *U.S. Department of Housing and Urban Development*
Kenneth J. Saulter *U.S. Department of Energy*
Roger W. Schmenner *Duke University*
Robert Schmitt *National Association of Home Builders*
Robert J. Sheehan *National Association of Home Builders*
Joseph Sherman *U.S. Department of Housing and Urban Development*
Kenneth A. Small *Princeton University*
Edward J. Smith *U.S. Department of Agriculture*
Raymond J. Struyk *Urban Institute*
Howard J. Sumka *U.S. Department of Housing and Urban Development*
Grant P. Thompson *Conservation Foundation*
David L. Weimer *University of Rochester*
Michael C. Wells *U.S. Department of Housing and Urban Development*
William P. White III *U.S. Department of Energy*

Index